MEASURISM

Hiwa Modarresi

PREFACE

The aim of this book is to discuss Physics from a religious perspective and to provide a solution to its fundamental misconceptions in a way that is understandable to all readers. In line with this approach, it is tried to describe various scientific phenomena down to a comprehensible level. Considering the extent of the scientific material covered, however, some readers might find it helpful to obtain further information from other sources with regards to certain topics.

This book starts with a chapter discussing reasons that make brains follow a logical or religious trend when processing certain topics. In the second chapter, most significant reasons that place Physics among religious ideologies are discussed. The third chapter presents a measuristic approach in addressing scientific challenges of Universe that attributes a measurable character to every single material property in Universe. At the end, an annex is provided in which most fundamental scientific phenomena are explained based on the presented measuristic approach.

CHAPTER ONE: RATIONALE

MEMORY NODES

A child may remember the meaning of number *four*, for example, only by counting all the way from a lower number for which the meaning is already known to the child. The counting process at early age is often helped by finger gestures in which each number is associated with a specific gesture of stretched-out fingers. The phrase or word that triggers the counting process is probably the most fundamental piece of biological memory around which the entire numbering logic in a child's brain is going to be built. Once stimulated to count due to internal brain activity or externally after being asked to *count*, for instance, a child would most likely start counting from either *zero* or *one*. These starting points are fundamental brain events that are essential in the counting process. Other numbers are then established in reference to either of the two in a sequential chain of events. For those children who start the counting process while looking at their closed fist, the fundamental event in the counting chain is probably number *zero* or *nothing*. Alternatively, in children who start the counting with one of their fingers stretched, the fundamental event is recorded as number *one*. In both cases, the triggering of the fundamental event helps the children initiate the counting process and arrive at the number where they have already mastered, event-by-event. In other words, a child may remember the meaning of number *four* only after remembering the meaning of number *three* which is on its own meaningful to the brain if the meanings of numbers *two* and *one* are previously evoked in a successful manner.

The sequential processes such as counting or making short sentences are among the earliest recipes formed in human brain. Within this context, every sequential process which is developed by brain to handle one specific topic is referred as a *nodal recipe*. Nodal recipes enable brain to arrive at any destination *informational node* by evoking the nodes revealing the path towards it. Each informational node among the sequential nodes, forming a nodal recipe, is evoked only after its previous node is evoked. From this standpoint, each individual number within the counting recipe is a node storing a single characteristic of a specific number. A relevant example of such nodal recipes is where a passphrase is stored in brain. Most of the time, the whole passphrase can smoothly be recalled once the starting node of the passphrase recipe is triggered. That means, even if the brain wants to only remember the last character of the passphrase, it should always start by remembering the beginning of the passphrase.

Unless evoked by one of the senses, the remembering process in brain always happens in a sequential manner. The process of arriving at any of the nodes within the brain occurs in two different ways. In the first approach, as in the case of the numbering sequence, brain evokes events through the underlying nodes node-by-node until it arrives at the desired node. In the second approach, to arrive at a destination node, one of the underlying nodes is tagged with an address-link to the destination node containing the relevant information. The latter process of arriving at destination nodes is necessary to avoid redundant data storage by the brain. It is worth mentioning that tagging process is not a mutual process between the two connected nodes. That is why seeing a familiar-looking individual where the person is supposed to be stimulates a complete chain of node-triggering events from place-node to person's identity-node, while seeing the same familiar-looking person at a different place might render the brain unable to track back the place where the person belongs. This is because the person's whereabouts node is

3

tagged with an address-link to person's appearance and identity while the node storing the person's appearance or manners is not tagged with an address-link about person's location. In such a situation, the activation of nodes with no address-links to lost nodes usually renders the brain stuck at a wrong node and prohibits the evocation of correct nodes containing the desired information. However, as soon as a clue is passed over to the wondered brain, through one of the senses like hearing or seeing, a separate node is accordingly activated which might enable the brain to arrive at the desired missing place-node.

Brain's tendency to recall nodes in a sequential manner is a direct consequence of Brain's sequential pattern of memorization process. The sequential memorization is the fastest and easiest way of learning since incoming information is being attached to an already established node. For instance, consider learning a new word. If the new word is visually, phonetically, or through its meaning is somehow identical to an already learned word, then it is most likely stored as an extension to the existing word node. This implies that remembering the newly stored word would inevitably results in remembering the similar underlying word. Through practice, however, more nodes are tagged with address-links to the newly learned-word node that may facilitate its accessibility over time. After this, evoking the newly learned-word node may no longer trigger the remembering of the old underlying word.

If brain finds no similarity with words within its existing vocabulary inventory, the new word node would possibly be developed with a direct link to the recipient sensory organ which in this case will probably be seeing or hearing. This way of learning would generally lead to a much faster utilization of the word since the remembering process does not involve evoking unnecessary nodes. However, since no other established node helps in remembering and therefore nurturing the newly learned-word, the new word is much more prone to be forgotten. This way of learning will usually require more time as the brain needs more cellular activity and resources to establish and tag relevant nodes with address to the newly stored word.

Involvement of more senses increases the learning efficiency since it results in a larger cellular activity and thus more resources for the newly established node through an increase in number of poking incidences. It is worth mentioning that memorization process, especially through different senses, might in some cases lead to the formation of separate nodes for the same concept which may form the foundations of independent sequential nodes. That is why hearing a specific word may cause recalling different events than reading its equivalent.

Traceable nodes within a brain can be evoked either directly through the senses or after being addressed by other nodes. The latter way is essential in enabling brain to function and reason without being distracted by the sensory organs. This is equally important in bringing prominence to nodes that were originally founded on other nodes. For instance, although at early stage any of individual number-nodes are merely accessible via number-nodes located earlier within the counting recipe chain, in experienced brains all single digit numbers form independent nodes that are directly accessible by nodes from outside the original recipe.

Much of the data received by brain are stored with very faint cellular marks. This includes a significant portion of the incoming daily aural and visual data that are subsequently forgotten or overwritten. The loss of data in brain is not specific to nodes with faint cellular marks. Even nodes which are stored with strong cellular marks, if not revisited regularly, at some point will be fully or partially overwritten by neighboring cellular activity. A node whose address is tagged by many other nodes, and is therefore constantly revisited, is immune to annihilation. On the other hand, a node may lose its connection to underlying nodes or be partially or fully overwritten over time if not revisited in time. Examples of these situations are seen in the subject materials learned years ago from which only general highlights are excerpted in the form of newly established recipes and informational nodes. For such subjects, many of the original nodes are partially or totally overwritten and many more have lost their address-links to relevant nodes. In general, the longer the nodes are not revisited, the higher chances are that they are forever forgotten.

Recipes prepared by brain regarding different subjects decide on how effective the brain is while dealing with those specific subjects. These recipes are sieves and definitions that are developed to filter incoming information and to establish the most appropriate responses. The more accurate and detailed recipes within the brain are, the more reliable its verdict after the analysis of the topic would be. An expert brain with regards to a particular topic, for example, holds many recipes and nodes about the topic that take part in analyzing different aspects of the topic within its reach of knowledge.

From a biological perspective, every node is a unique constellation of cells. Nodes are mainly formed in response to the external stimuli received through one of the input sensory organs or the internal stimuli which are prepared to trigger the output organs like speaking or limb movement. Nodes can also be formed because of brain's own reasoning after some of the existing recipes and nodes are screened by other recipes. Generally speaking, brain is a collection of intertwined nodes that are tagged with addresses to relevant nodes and in which every node is able to replicate its perceived information upon evoking.

The most significant issues concerning the data storing or restructuring by brain are 'how much of nutrition is available to the brain cells' and 'how focused the brain is' the moment it has received or analyzed the information. A combination of satisfactory nutrition and interested brain facilitates the creation of new nodes or modification of existing ones. However, most of the time these conditions can only be maintained for a short period of time which renders the cellular adaptation and therefore the learning process generally a slow trend. This necessitates practice or in better words revisiting established nodes over and over again to re-nourish the memory cell constellation and make intended transformations. It is worth mentioning each time a node is revisited, whether it is aimed at storing or recalling information, the cellular constellation of the node structure would at least slightly be transformed. The transformation might happen for different reasons but mostly occurs to adapt to the reason that has caused the revisit.

RELIGION

The slow process of learning makes the environment by far the most effective means with regards to node formation. That is how a specific habit, culture, belief, or religion is established in brain. An absolute majority of religious people have come to believe in the religion they follow only by strengthening the nodes which had been planted in their brains since the times they hardly remember. In fact, long before a brain is sufficiently experienced to question the surroundings, nodes answering those questions have already been formed in line with the governing environmental circumstances.

The main reason making people create and embrace religion lies in their need to find answers to questions that lack conclusive answers. The most notable of such questions are 'where have we begun', 'where are we heading', 'what for', plus a whole range of other related *why*, *who*, *what*, and *when* questions. Religion can best be defined as any system of members who work actively or passively to unify their ideology regarding unanswerable questions. In extreme cases, the unification of ideology extends to include questions that may have definitive answers.

Religion, as established inside the brain, is for most part a collection of inter-tagged recipes that adapt the incoming information to the existing nodal structure or label them as false if they are found not to comply with the conditions required by at least one of the existing recipes. Usually, besides the nodes created as conclusive answers to the unanswerable questions, religion interferes with the formation of plenty of nodes covering all sorts of topics inside the brain of constituent members. Some of these affected nodes include routine daily practices which are tagged with correct and incorrect keywords or are carefully transformed in a way that are deemed to best serve the religious ideology. As time goes by, the network of religious nodes and their influence in decision making within brain grows larger which makes it increasingly difficult to convince the brain of its misconceptions.

Religious brains tend not to reexamine the credibility of the developed recipes or nodes when revisiting them for at least one of two main reasons. One reason is that brain itself prevents any reexamination because it contains a recipe that prohibits questioning the credibility of other recipes or nodes deemed as utterly correct or sacred. The alternative reason is that brain might give up any reassessment since it would take an enormous amount of concentration and nutrition for it to suppress or transform any of established religious recipes or nodes and to consequently readjust all related recipes and nodes, while in the meantime maintaining brain's desired consistency level.

Majority of members in a religious system accept a certain practice or belief as correct or incorrect solely because others in the system consider it as such. This way of judgment encourages the religious mindsets to mostly outsource the decision making to others and convince themselves that some people within the religious system know the complete answers to the unknown. Yet, in fact, nobody knows the exact answers to the unknown. Should the answers had been known, the religion would have not existed at first place.

LOGIC

Whether created or revisited, the nodes are always checked for consistency and are mostly formed or reformed to support the general trend followed by existing nodes. The constant analysis of incoming information for consistency is performed by existing recipes. Normally, there is no strong reason for a brain to reevaluate the authenticity of existing recipes or nodes. However, there are certain situations which may force the brain to consider reassessing the validity of existing recipes or nodes. If needed, such reevaluations are more difficult for those recipes or nodes on which a growing number of other recipes and nodes rely for credibility. A brain which is willing to reevaluate every existing recipe or node within a specific topic is considered to have a logical approach within the framework of that topic.

The most serious test that reveals the religious or logical approach of a brain towards a topic occurs when conflicting information with high veracity is received by the brain. In a religious approach, brain labels the newly received information as false because it is unwilling to question the authenticity of the existing foundational recipes and nodes within the topic. On the other hand, a logical approach of the brain tends to evaluate the validity of the incoming information while at the same time slowly reevaluating the authenticity of the existing recipes and nodes. Unlike a religious brain, a logical brain sees a *claim* far from being a *fact* which needs to be supported by at least one piece of *evidence*. According to its final verdict on the issue, the logical brain may even discredit its most well-established recipes and nodes and start the tiresome process of restructuring a whole network of recipes and nodes within the topic. Such restructurings usually last for long durations of time as the growth of new recipes and nodes as well as scrutinizing process of all the relevant nodal structures and their subsequent ramifications need significant level of focus and nourishment.

The tedious procedure of reevaluating the existing recipes and nodes may cause challenging times for brains with logical approaches, at least during early stages. However, after this difficult initial period, information within such brains would be much better structured and supported by consistent recipes and nodes that make the future analyses much faster and more consistent.

It is important to note that every brain starts from a blank stage and develops in line with the nodes being formed within it. Because nearly the entire initial nodes are formed due to environmental stimulations, all brains in their initial form have religious approaches towards various topics. It is only after receiving credible conflicting information by an experienced brain that its logical character might be surfaced. A brain that has never received any reliable conflicting information has in fact had no reason to transform itself from religious to logical with regards to a certain topic.

CHAPTER TWO: PHYSICS

RELIGION OF PHYSICS

For thousands of years religions have evolved according to surrounding circumstances and managed to maintain prominence. Every religion proclaims itself as the unique way towards a salvation which is primarily meaningful within the framework of assumptions taken by the very same religion. The most fundamental aspect among all religions is the outsourcing of rational by the constituent members to the religion about what is correct and what is incorrect. In this regard, Physics is no stranger to the world of traditional religions and although it claims to merely be an outcome of scientific logic, its fundamental structure is barely different. Considering the mainstream scientific approach pursued by the majority of physicists as well as the claims made by many of Physics' revered elites about proving certain aspects of traditional religions, overall, it is fair to say that today's Physics is the new millennium religion at its best and a continuation of any other religion at its worst. To clarify the religious character of Physics, its general similarities with other religions are highlighted here.

1. *PHYSICS HOLDS THE IDEA THAT THERE HAS BEEN A START TO UNIVERSE AND THAT THERE WILL BE AN END TO IT.*

Physics believes in a *start* for Universe and relates its creation to an explosion it terms as Big Bang. Although some physicists undermine the significance of Big Bang by emphasizing its theoretical aspect, its widespread acceptance and the numerous offspring theories emanated from it has already rendered it a fact within Physics' community.

Besides conceiving a start, Physics also believes in an *end* to Universe. Even though Physics lacks a clear stance on the final format of its end story for Universe, its current principles anticipate a freezing end to it. As deemed by Physics, the freezing will be facilitated through the expansion of Universe known as Expansion theory which itself is one of the fallouts of Big Bang theory. According to this perspective, Universe is aging towards a zero Kelvin era in which all the energy is radiated away from matter. Obviously, the presumed freezing doomsday scenario predicted by Physics will only be materialized if planet Earth were to survive being consumed by any of Universe's imagined *black holes*.

2. CONSIDERING EARTH TO BE AT THE CENTER OF UNIVERSE.

Because regardless of their direction, farthest observable objects in the sky are located at comparable distances from Earth, to physicists, planet Earth is apparently believed to be at the center of Universe. In this respect, the borders of Universe stand at where the farthest bright objects of the sky have supposedly reached following the Big Bang explosion.

According to Physics the borders of Universe are located not farther than a distance of around tens of billions of *light years*. It is important to recall that the primary method in determining the distances of farthest galaxies from Earth utilizes the occurrence of *red-shifts* within the spectra collected from these galaxies which on its own renders the estimated distance of these galaxies highly questionable. A red-shift is referred to a displacement of spectral lines corresponding to an element towards lower frequencies. Physicists interpret the occurrence of these red-shifts within the Expansion theory according to which the farther objects, regardless of their location with respect to Earth, always exhibit larger red-shifts. Based on the current astronomical observations and assumptions, planet Earth is almost exactly located where the assumed Big Bang explosion has occurred. Thus, according to physicists, Earth has been and will always remain at the center of Universe.

3. *PRESENTING ITS CLAIMS AS FACTS.*

Physics presents all its claimed laws, rules, postulates, theories, and principles as ratified facts even though majority of them can never be backed by scientific evidence. Physics is sturdily intolerant of critical questioning of any of its assumptions which is ironically in stark contrast with its profound desire in portraying a tolerant and rational image of itself as is expected from an ideology capable of thoroughly addressing the fundamental issues.

4. *JUSTIFYING WHATEVER PHENOMENON ONE WAY OR ANOTHER AS CONDITIONS DEMAND.*

Physics presents explanations to the experimental phenomena in whatever way that serves its interests the best. With this attitude and in the absence of other viable justifications, Physics has presented some of the most ridiculous explanations regarding certain phenomena. The most interesting of such explanations have been presented after new experimental findings have created awkward situations for Physics in which the credibility of one or more of its established principles have been threatened. An example of such a situation has occurred around the well-established Rayleigh limit, also known as *angular resolution*, which sets a limit on the resolvable distance between two bright points based on the wavelength of the light. When later in a technique known as *near field optical microscopy* it was discovered that the limit of resolvable distance could in practice be much smaller than what was stated by the Rayleigh limit, Physics explained the reason by introducing a new concept of *non-propagating evanescent field*. It is, however, impossible to prove or disapprove the presence of the mysterious evanescent field or its non-propagating nature as it is supposedly not propagating to start with, contrary to the concept of conventional electromagnetic waves. Furthermore, Physics doesn't seem to bother with explaining how photons make the journey to the eyepiece lens in such experiments considering the non-propagating nature of the waves involved. Much of the ample maneuverability space around Physics in explaining various phenomena, and as it pleases, is due to the vague nature of its justifications given to underlying scientific issues.

5. *INSTRUCTING NUMEROUS EXCEPTIONS ALONGSIDE MOST OF ITS DEVISED PRINCIPLES.*

Despite Physics desire in portraying itself as a highly well-structured ideology, there are extensive caveats inherent to its structure requiring the introduction of exceptions. These exceptions are mainly there to serve the short-term interests of Physics and therefore new exceptions are constantly devised to the recently risen problems driving the succession of ill-advised solutions until eternity. This behavior is essential to help the religious establishment appear potent in the eyes of its constituent members. There are hardly any laws in modern era Physics without exceptions. In most cases, the exceptions themselves are wittingly introduced as independent laws and rules of their own. Take for example, the Selection rules that are arbitrarily applied to various electromagnetic interactions or all the rules that are imposed to sustain *Rutherford-Bohr* atomic model.

6. *MANY OF ITS FUNDAMENTAL PRINCIPLES CONTRADICT ONE ANOTHER.*

Physics' inability to admit and consequently correct its earlier mistakes is the reason why some of its principles will always remain in contradiction. Admitting mistakes is considered a weakness and therefore despised by Physics. Instead, Physics claims all its rules are there for good reasons. What makes the situation even worse is Physics insistence on coming up with answers to newly raised questions as quickly as possible hoping to further present itself as a highly capable ideology. This leads to the creation of short-term solutions which over time may turn contradicting solutions required for future findings. A remarkable example of a conceptual contradiction in Physics is the claim of simultaneous current flow and exclusion of magnetic field inside superconducting materials. As it is known in today's Physics, zero-resistance of superconducting materials against the current of electrons has enabled scientists to produce very stable magnetic fields. In practice, highly stable magnetic fields generated by superconducting materials are extensively used in various medical and characterization equipment. Physics, at the same time, has another principle regarding the superconductors which is in stark contradiction with the previous one. This principle is known as Meissner effect and states that any external magnetic field is expelled from superconducting materials as they cool down below their superconducting temperatures. Thus, according to Physics, it is a lossless current of electrons inside a superconducting material that creates the extremely stable external magnetic field while at the same time there can be no magnetic fields inside the superconducting material as explained by Meissner principle. For Physicists it seems the superconducting material is somehow able to selectively identify and let the internally generated magnetic lines pass through while expelling any magnetic field lines generated by external sources.

7. *PRONE TO BE ABUSIVE.*

The abuse in Physics happens both systematically and on individual levels. A grave example of systematical abuse in Physics is the case of many prestigious prizes won by the so-called physicists whose only contribution to the award-winning research has been their authority over the work of genius researchers. On individual level, physicists can exploit the ones under their authority mainly because of a lack of independent supervision and due to the extreme powers given to the so-called elite within Physics. Whether the abuse is systematic or on individual level, it is Physics' establishment that tolerates and allows for such abuse. Most of the time, such abuses are easier seen from outside the religious system since the inner members have already accepted the abuses as norms or are totally overlooking them. In practice, the enslaving religious system is always appreciated by both the elite and low-rank followers.

Constituent members believe in religious system benefiting and raising only a limited number of devout members which leaves majority of the members in serving positions. In such systems, the so-called devout members are most of the time easily recognizable through the venerated titles prefixed to their names.

8. *DESPISING INDEPENDENT THINKING.*

Physics never negotiates on the authenticity of its assumed principles. All constituent members are expected to fully embrace these principles as the only true way towards understanding Universe. Thus, although Physics publicly pretends to be admiring critical thinking, it deems most of such attitudes as a disapproval directed at its establishment. Exceptionally, however, in cases where the new critical views somehow acknowledge some of the established principles within Physics, limited variations of such views are tolerated.

9. *TAKING CREDIT FOR WHATEVER CONSIDERED NOBLE.*

Physics directly or indirectly takes credit for all the human progress and whatever considered good, while at the same time laments other ideologies for causing backwardness. Physics tends to portray today's technological advances as achievements obtained mainly via its vision of Universe. Nevertheless, an absolute majority, if not all, of the technological breakthroughs so far have been direct consequences of experimental findings solely accomplished via accidental discoveries or trial-and-error. Despite, Physics' portrayal of itself as the ideology predicting and expecting those breakthroughs via consequential logic, in fact, its role has essentially been nothing but interfering after the experimental observation of every effect and presenting an explanation regarding the cause driving the phenomenon in such a way that suits its vision of Universe the best.

10. DEPICTING SEAMLESS IMAGE OF ITS ELITES.

Physics portrays its elite members and founders as the infallible icons who have immensely benefited the humanity through their ingenious contributions and that without their contribution it would have been impossible for humanity to be where it is now.

11. MAINLY FINANCED THROUGH DONATIONS.

Physics has promoted itself as one of the fundamental needs of humanity and has convinced various elements of the society to financially help it survive and prosper. At the same time, ironically, it has maintained authority over mechanisms issuing progress assessments required by some donating organizations for the extension of their financial support.

12. CONSIDERING THE SOCIETY MEMBERS AS RIGHTEOUS OR IGNORANT.

Physics considers those who believe in its ideology as the righteous members and those who do not fully embrace it as ignorant. Physics' power structure systematically obstructs the spreading of idea's owned by those it deems as ignorant.

13. *ITS TIME AND EFFORT IS MAINLY DRAINED BY DISCUSSIONS BORN OUT OF ITS OWN PRESUMPTIONS.*

Physics spends much of its time and efforts in clarifying topics which only exist due to its own earlier assumptions. Although Physics depicts such discussion as necessary, in fact they hardly have any constructive effect on human life or understanding of Universe. As an example, consider the transmission of light through a coated glass for which the refractive index of the coating is slightly less than that of the glass and that the thickness of the coating is comparable to the half wavelength of the incident light. This concept basically tries to relate the transmission of light through the coated glass to the *destructive superimposition* of light reflections from both the coating and glass. Based on Physics' wording, intensity of transmitted light through the coated glass is maximized because light reflection is minimized due to the destructive superimposition of reflected waves. In other words, annihilation of photons on one side of the coated glass supposedly leads to the generation of photons on the other side. In fact, nobody outside Physics is interested in comprehending such nonsense and the only useful takeaway message here would be that the coating on the glass determines its transmissive or reflective behavior during an encounter with a beam of light. Another relevant example concerns the effect of gravity on light. Since based on Physics' claim that energy propagation in the form of light involves no mass transfer, physicists should somehow find a way to justify why light beams coming from faraway galaxies are bent by gravity force such as what is seen in phenomena known as *gravitational lensing*. Instead of presenting a rational solution, however, Physics goes through the tiresome work of mathematical obfuscation of the problem and at the end claims that it has solved the mystery it created at first place by presenting what it calls curving of the *space-time*. Yet in fact, it is the initial erroneous assumption of zero-mass of photons which creates this whole need for hopeless follow-up problem obfuscations.

14. PRESENTING AN INTRICATE IMAGE OF UNIVERSE.

Physics suggests an intricate network of rules governing various interactions within Universe. This is, at least implicitly, to dissuade society members from expecting the details behind Physics' so-called principles and that only the elite physicists may know the complete answers. Yet, in fact, Physicists have no convincing answers to questions regarding fundamental phenomena in Universe.

Another benefit of presenting an intricate image of Universe is in creating loopholes for future maneuverability when it comes to new justifications required for vaguely known issues. It is fair to say that today's modern Physics is mainly the outcome of confusion created by Physics itself.

One of the primary mechanisms of making issues intricate is to use mathematical formulations involving complex numbers which can often be accompanied by twisted assumptions. Solutions derived from such formulae are most of the time non-unique and in practice, based on the path and assumptions taken, might even lead to rather contradictory conclusions.

15. *ADVOCATING SUPERNATURAL CONCEPTS.*

In line with portraying an intricate image of Universe and its governing rules, the supernatural concepts add to the intricacy that religion tries to associate with the philosophy of existence. Believing in supernatural concepts keeps the religious society always hopeful in fortunate coincidences or changes brought forward by the unknown. *Quantum entanglement* can be mentioned as one of the most remarkable supernatural concepts advocated by Physics. According to this concept, every independent particle such as an electron can travel in time to its past and re-adjust its state before presenting itself to an upcoming measurement that expects a specific state of the electron. Another remarkable supernatural example in today's Physics is the concept of *quantum tunneling*. Based on this concept, for every two pieces of matter there is always a chance that they pass through one other upon collision. The concept of *dark matter* makes another example of supernatural concepts advocated in today's Physics. This concept has been coined to justify the constant *angular speed* of stars located within the discs of spiral-galaxies. In fact, the number of supernatural, or better called superstitious, concepts in Physics is comparable with, if not more than, those in any other religion. Physics provides itself with some level of safety margins by labeling many of its superstitious ideas, as hypothetical. However, in practice Physics has for long embraced all its fundamental superstitious believes, plus a whole host of additional topics emanated from them, as indisputable facts and has been persistently trying to find ways to further consolidate their standing. In fact, the authenticity of today's Physics is tightly tied to the credibility of its ample superstitious concepts.

16. *UNDERMINING THE IMPORTANCE OF ADDRESSING PREVALENT CHALLENGES.*

Physics evades detailed discussion of even the most ubiquitous questions when it notices the potential dangers posed by the presence of logical-gaps within its presented justifications. It does so mainly through creating and magnifying fancy hypothetical challenges that often overshadow discussions encompassing the more fundamental ones. For instance, Physics is by far more involved in bragging about the consequences of Big Bang rather than talking about the likely conditions, if any, that could have led to such a presumed explosion.

FLAWED PRINCIPLES

Majority of the pressing phenomena shaping today's modern Physics were discovered in late nineteenth and early twentieth centuries. This period also witnessed significant advances in radio and televised communication technologies that facilitated the exposure of scientific debates to the public. At the time, the publicization of scientific challenges exerted extra pressure on Physics' community to come up with explanations regarding the unknown. To protect its position as a capable ideology, Physics hastily provided erroneous or at best twisted explanations to a great number of scientific challenges in a span of one decade, leaving hardly any fundamental issues to be resolved later when adequate information would have perhaps become available.

Today's Physics follows the same path taken primarily by the recognition of Rutherford-Bohr's atomic model in 1913. The slow introduction of the *relativistic theory* during the coming years followed by erroneous explanations of the underlying experiments of

Young and Michelson-Morley, all in the absence of viable alternative scientific explanations, led Physics towards the path it is still pursuing. Aggravating the situation was the new adaptations to Relativity theory that despite its highly questionable assumptions only received criticism during its first years of introduction. Since then, the Relativity theory has turned into the fanciest theory of all time, ostensibly understood only by the *wise*. It is extremely unfortunate that most of Relativity-theory sceptics are reluctant to express their frustration out of reservations of being labeled as not intelligent enough to understand the theory.

The harmonization of all the erroneous principles and twisted interpretations of Physics started nearly a decade later by the sluggish and everlasting introduction of Quantum Mechanics which was primarily established following the works of Heisenberg and Schrödinger. Quantum Mechanics along with its offspring Quantum Electrodynamics and Quantum Chromodynamics have since been utilized as an affordable filling substance to obscure the wide gaps constantly emerging from forcefully reconciling the numerous contradictory principles of Physics.

Physics' inherent network of complex and intertwined paradoxical justifications has made it very difficult to bring corrections to its structure. Its ultra-complex establishment has successfully shielded it against all possible criticisms even regarding some of its most insane principles. Probably the best well-received example of such a case is the simultaneous application of wavelike and particle-like behaviors to light and other fundamental particles. Physics arbitrarily uses either of these doubly assigned characters to justify the observed effects as they better suit its justifications. In line with this senseless approach, Physics has introduced yet another principle known as *measurement problem* which basically states that a particle may only appear in a particular place when it is measured. In other words, according to measurement problem, at the same time both there are

23

and there are no particles everywhere, or equally nowhere. This absurd conundrum is better known as *Schrödinger cat* and provides a perfect platform to further some of the most ludicrous theories of Physics. An exemplary circumstance that clearly characterizes the Schrödinger cat conundrum is the single photon experiment in Young's interferometer. Based on Physics explanation of the interference theory, the bright and dark fringes are formed in the interferometer even when only one photon is shone to the slits. According to the Schrödinger cat principle, however, as soon as a parallel photon-detection-measurement is performed to validate the claim, it becomes evident that single photon does not lead to fringe formation because the only photon apparently always shows up where the photon-detection-measurement inside the interferometer is performed. The ridiculous outcome of this justification is that Physics safely retains its claim that even one photon leads to diffraction pattern formation which is a necessary requirement for its earlier claim of the wavelike behavior of light, without having a slightest worry about being proven wrong. Within the same context of paradoxical theories of Physics, another remarkable example is the black hole fallacy. According to Physics, black holes are so massive that nothing can escape their gravity force even the light. This implies that apart from darkness, there will be no observable record of any black holes ever, and therefore it will never be possible to validate or invalidate its existence. It is worth mentioning that the empty space observed at the center of many galaxies is just a vortex center void of matter like any other vortex center seen for example at the centers of whirlpools and whirlwinds and has nothing to do with a black hole. Galaxies are not stabilized because of the assumed existence of black holes at their centers, but because of the balancing gravity-force created by matter dynamics on the opposite sides of every galaxy's center of mass. Similar invalidatable paradoxical subjects are widespread over various fields of Physics. At the same time, validating any of its senseless effects, theories, or principles is not among the concerns of Physics since they are expected to be known as facts anyway, especially in the absence of any presented rival explanations.

There are certain historical events that have for most part shaped Physics as it is today. These fundamental events are discussed here.

AETHER ILLUSION

Aether can probably be named the most underlying erroneous principle forming today's Physics. When early physicists tried to explain the mechanism of light propagation in outer space, they thought it was necessary to have an invisible grid-like medium using which light could propagate. They subsequently called this imaginary medium Aether and assigned it with a character similar to air which was the most obvious medium using which sound could propagate.

A fixed speed was at first assigned to light travel based on an analogy to the speed of sound travel in air. The speed of sound or any other oscillatory spreading of kinetic energy only depends on the characteristics of the medium through which the wave propagates. Based on this analogy, light was assumed to travel with a constant speed, merely decided by the characteristics of Aether medium. Although, later it was claimed that the concept of Aether had been refuted by the outcome of Michelson-Morley's experiment, in practice the erroneous interpretation of the experiment not only approved the concept of Aether but took it a step further. Currently Physics considers the speed of light propagation to be a fixed value regardless of the reference coordinate system used in determining the speed. That means, in simple words, the speed difference between every two neighboring photons within the same traveling beam of light would always miraculously be equal to the speed of light propagation.

Another underlying problem forming today's Physics is the concept of wavelike behavior which was first applied to the light propagation and later extended to other particles. Although the wavelike behavior of light had partially been accepted through resembling its propagation in Aether to that of sound in ambient media, its proof of concept came only through interference patterns observed in Young experiment at the start of nineteenth century. In Young's experimental setup, bright and dark fringes of light are created on a screen when light is shone on it from a distance after passing through two closely placed narrow slits. At the time, the fringe formation for physicists resembled the interference patterns observed among water waves inside a ripple tank. The Young interference experiment since then has been the strongest proposed proof of the wavelike behavior of light.

To fully account for a wavelike behavior of light, however, it was necessary to explain the interaction mechanism between the waves passing through the two adjacent slits in Young's interferometer setup. Therefore, an additional justification was needed to bring an exception to the overwhelmingly accepted fact that light was travelling in straight lines with little divergence. The exception should have stated that light propagated in wide angles after passing through each slit. Physicists at the time backed their imaginations using the notes of Huygens who wrote about a similar idea nearly a century earlier. Concurrent with Young's studies, Fresnel adapted Huygens' idea to formulate his own mechanism of diffraction pattern formation in what is today known as Huygens–Fresnel principle. The adopted principle states that every point on a wave-front might be considered a source of secondary spherical wavelets spreading in forward direction at the speed of light and that the new wave-front is the tangential surface to all the secondary wavelets. To a non-physicist rational this implies as expecting every point on the beam front possessing intelligence which enables it in deciding and generating a tangential secondary spherical light wave spreading only in the direction of originally propagating beam and which only occurs after light beam passes through small openings.

Physics takes the wavelike behavior of matter to an extra level of absurdity by applying the same wavelike concept to single particles such as electrons when shone on slits in Young's experiment. As explained by Physics, in a phenomenon termed as *electron wave enigma*, closely entangled with the Schrödinger cat conundrum, as soon as the position of an electron on its way towards the interference screen is known, a time travel of electron to its past occurs. The miraculous time reversal takes the observed electron all the way back to where it earlier entered the slits as wave and subsequently transforms itself into a particle before re-entering only to the slit where the electron detection was performed. The re-entered electron, at zero time difference meets the detection device and this time reveals itself as a particle. In other words, for physicists, a single electron can cause the creation of bright and dark fringes once it is shone on slits of Young's interferometer since the single electron enters the setup as a wave. However, once a parallel experiment tries to establish the wave character of the shone electron, the electron resists giving away its wavelike identity. Therefore, the electron travels back in time and re-enters the setup through the slit on which the parallel measurement is performed and identifies itself as a particle. For physicists, the observation of electrons as particles on their way towards the screen renders the interference pattern formation unsuccessful, as electrons are expected to be waves within the interferometer apparatus. However, the truth is that by measuring the presence of electrons at the position of one of the slits, electrons are blocked from traveling the rest of their trajectory to the screen. Therefore, fringes are not formed only because electrons on one of the beams are blocked by the detection apparatus. Yet, physicists are astounded by the absence of diffraction pattern on interferometer screen once a parallel measurement is performed to establish the wavelike behavior of electrons.

Instead of admitting the particle nature of electrons as it is partly evidenced from the outcome of Young's experiment, physicists further misuse their false justification of the Young experiment to bring credibility to yet another absurd theory of theirs which is known as *uncertainty principle*. Uncertainty principle is widely used by physicists to attribute dual wave-particle behavior to all moving matter. However, at its root, the uncertainty principle drastically fails in establishing a consistent reference coordinate system regarding its definition of speed.

Around the beginning of twentieth century, Physics obtained all the extra evidence it needed to accept the sheer particle-like character of light through Compton and Photoelectric effects which clearly demonstrated the interaction between light and electron particles. Unfortunately, nonetheless, the strong evidence provided by Compton and Photoelectric effects was apparently not clear enough to compel Physics into reconsidering the wavelike character it had assigned to light and other fundamental particles. The final verdict of Physics was the assignment of both wavelike and particle-like behaviors as conditions demanded. Since then, Physics has constantly tried to obscure the ever-emerging fallouts of the dual wavelike and particle-like behavior-assignments to light using complexities provided by Quantum Mechanics and its offspring branches.

CONSTANT SPEED OF LIGHT

In the late nineteenth century Michelson and Morley performed their famous experiment, named after them as Michelson-Morley, which followed long debates on Aether and light propagation mechanism. Perhaps one of the most fundamental mistakes that has led Physics to be where it is today was the failure of Michelson, Morley, and other physicists of the time in correctly interpreting the fringe formation mechanism in Michelson-Morley's interferometer apparatus. The psychological environment that Aether and wavelike behavior of light had created within the scientific society during those days played a significant part in the final adoption of their mistaken interpretation. According to Michelson and Morley, changes in fringe layouts could have only been occurred due to interference between two light rays with slightly shifted wave patterns. They expected a small shift in wave patterns of the two branched light rays in their setup to be caused by the difference in the speed of light as measured relative to Earth's direction of motion in Solar system. The fact that Michelson-Morley's interferometer showed no changes in the layout of fringes, for measurements performed during different seasons, made them and other physicists erroneously interpret the outcome of the experiment as a proof that light was traveling with a constant speed in all spatial directions irrespective of the relative motion of Earth around Sun and through Aether.

Interestingly, the concept of a constant speed of light travel was adopted by Physics' community at a time when a similar but more ingenious experiment clearly refuted the conclusions inferred from Michelson-Morley's experiment. That experiment is known after its designer, Sagnac, who arranged the same interferometer setup used by Michelson and Morley on a rotating table while mounting the light source on the same rotating table. Thanks to Sagnac's unique setup configuration, the effect of Earth movement in space could have easily been discarded. Sagnac's interferometer clearly showed varying fringe formations by changing the setup's rotation speed or direction. Based on the outcome of Sagnac's interferometer, light speeds of the two branched beams were differing according to the dynamics of the rotating table which consequently resulted in changes to fringe layouts, the same change Michelson and Morley wanted to see but couldn't owing to their setup's design limitations. Unfortunately, even though the variations of the Sagnac interferometer since then have widely been used for precise gyroscopic navigation purposes, instead of cherishing the experiment and utilizing it as a base for development of a suitable approach towards the issue of light propagation, the conclusions drawn from Sagnac's interferometer were silenced and any mentions of the setup were left out of educational curriculum to further portray the interpretations inferred from Michelson-Morley's experiment as unchallenged.

Nowadays, regrettably, the constant speed of light is one of Physics' most foundational principles and its credibility is closely tied to that of Physics itself. Despite the erroneous constant-speed-of-light conclusion drawn from Michelson-Morley's experiment, in fact, this and the more sophisticated Sagnac interferometers were perfect experiments that demonstrated the particle-like behavior of light. In these experiments, it is clearly seen how light beam undergoes speed changes as it reflects from moving surfaces. The reflected light beam gains an extra speed equal to the speed of moving mirror in case the mirror is moving against the direction of the incoming light beam. On the other hand, the light beam loses its speed by a value equal to the speed of receding mirror that moves in the same direction as the incoming light. In other words, the reflected light beam is blue-shifted once the mirror is moving against the direction of the light beam, while it is red-shifted when the mirror is moving away from the light beam. In Michelson-Morley's interferometer, once the branched light beams from the two perpendicularly placed mirrors rejoin at the position of the splitter, both of which have the same absolute speed values as they had during the splitting. This is because either of the perpendicularly diverted beams has had to make an even number of reflections and therefore an even number of speed-gains and speed-losses before arriving back at the splitter position. Hence, considering a fixed travel distance for either of the branched beams in the setup, it always takes the same amount of time for either of the branched light beams to come back to the point at which the original light beam was split, regardless of the dynamics of planet Earth within Universe. On the other hand, in Sagnac's interferometer, the speed of light beams between two perpendicularly placed mirrors differs according to the rotational dynamics of the interferometer setup. While a non-rotating Sagnac interferometer setup results in a similar outcome as observed in Michelson-Morley's interferometer, a rotating one introduces a sideward blue-shift or red-shift to the branched rays of reflecting beams according to the setup's angular speed and rotation direction.

A close look at all types of interferometer setups, reveals two essential conditions that must be met for interference patterns to form. One of the conditions is realized by employing the same source of light to create the branched light beams. This results in identical coherence pattern of the interacting light beams within interferometer setups. The second condition is achieved by superimposing the branched light rays at grazing angles. In general, whenever two or more beams of light with similar coherency meet under grazing angles, whether it happens in Young, Michelson-Morley, Sagnac, or any other interferometer setup, a discrete pattern of bright and dark fringes is formed at any distance on a screen blocking the propagation of superimposed beams.

ATOMIC MODEL

Absence of a viable atomic model that could explain the neutrality of matter while at the same time incorporated the known charged subatomic particles of the time, namely negatively charged electron and positively charged nucleus, can be mentioned as another pressing issue challenging physicists' knowledge in justifying scientific phenomena during early twentieth century. Among the numerous proposed models that were presented by different scientific circles of the time, Bohr's model was especially selected as it could acceptably be manipulated to justify the interaction of light with matter. In this model, electrons are assumed to rotate around the positively charged nucleus in a manner similar to the rotation of planets around Sun. Interestingly, Bohr's Solar-system-based atomic model was proposed in an era when the dynamics and characteristics of subatomic particles were still the subjects of heated debate. In fact, Bohr's atomic model was accepted only few years following the discovery of positively charged nucleus by Rutherford in 1911, less than two decades after the discovery of negatively charged electrons by Thomson, and especially around two decades before the discovery of neutrons by Chadwick. Both prior and after the adoption of Bohr's atomic model, many complementary laws had to be coined and applied to force consistency on its inconsistent assumptions. One of these laws became known as *quantum jumps* and was necessary to justify the light emission and absorption processes occurring in atoms after electrons assumingly jumped to different permitted energy levels. It was also assumed that the rotational orbitals of electrons were restricted to certain radii around the nucleus. This latter law was necessary to silence any critical argument about why the negatively charged electrons would not fall on the positively charged nucleus, particularly considering the constant energy loss of electrons due to their continuous electrostatic friction against other passing-by electrons. Furthermore, another law was required to prevent coexistence of any two identically characterized electrons within the same orbital trajectory. This is known as Pauli exclusion law and was essential to justify the multiple valence numbers shown by single atoms. Two decades later another additional assumptive law was imposed on the atomic model under the name of *strong nuclear force* that by the time became necessary to justify the extremely close-distance coexistence of positively charged protons within the nucleus of atoms other than hydrogen. The envisaged nuclear force was assumed to act and overwhelm Coulomb repulsion only in situations where protons were extremely close to one another. The proposition of strong nuclear force coincided with the discovery of neutrons which together with protons were assumed to form the nucleus and therefore the total measurable mass of atoms. Around the same time as the proposition of strong nuclear force, the fallouts from Bohr's atomic model continued even further with the introduction of *weak nuclear force* that suddenly became vital in creating a pathway to justify the radioactive decay phenomena shown by certain atoms.

TIME

Time indicates a rate of change between new and old states of any of matter's dynamic representations including its properties and location. Due to its comparative essence, the notion of time is only relevant when an observer tries to extrapolate the occurrence of a specific change-event. For example, the concept of change-rates as quantified by two independent fictional observers who can see either a rock located in an isolated corner of Universe or an isolated photon far from all other matter could be similar as both observers are restricted to quantify changes with respect to a single reference object. The rates of change, however, will be immensely different once a single observer quantifies the changes for both systems because in this case it would be possible to compare the rates of change in either of the systems to the other one. Thus, the definition of the term *time* is a relative concept that varies according to the presumed reference events and that association of fast-forwarding, slowing down, stopping, and reversing to the concept of *time* is absurd.

Binding *time* to position within the concept of Relativity theory in what is today known as *space-time* has contributed to numerous ridiculous theories negating even the *causality law* that states no event may precede its cause. Among these illusionary theories, the *gravitational waves*, which are assumed to spread with the speed of light, and *time mirrors*, according to which there are parallel Universes where time passes in reverse direction from the future to the past, are among prime examples.

MATHEMATICS

Mathematics in general is the most logical language ever developed by human being. Its regulations are defined within clear and rigid borders that always, regardless of the path taken, direct every problem towards a unique answer or set of answers. The logical framework of mathematics compels everyone to accept its derived solution for every single problem. However, the formulation of the problem itself can be a subject of scrutiny. It is the human brain who should be directing the mathematics towards the right direction through meaningful formulation of problems and not the other way around. Take 'zero to the power of zero equals one' in mathematics for instance. This mathematical law is there to get our definitions right and to direct the solutions towards the right answer. However, materializing such a mathematical regulation in real world leads to fallacy. This valid mathematical regulation in real world will read as 'creating one thing from raising nothing to the power of nothing'. Such twisted interpretations of mathematically driven formulations have already resulted in adoption of many illusionary theories in Physics. An example of this is physicists' belief that energy could be extracted from vacuum based on the illusion given by what is known as *zero-point energy*.

Similar misconceptions in Physics are driven by mathematics' utilization of its powerful *imaginary numbers* tool. 'The square root of negative one' is only there to help with finding solutions to various polynomial and differential mathematical equations. In real world, on the other hand, 'negative value of something' is on its own meaningless let alone materializing a 'something that results in negative of something if multiplied by itself'. Therefore, outmost care must be taken when applying some of the mathematical rules to real world trends and phenomena.

The wrong formulation of original problems along with uncontained interpretations of some of the mathematical outcomes, especially in topics related to Quantum Mechanics, has made Physics very susceptible to the emergence of superstitious theories. Illusions defined within the framework of what is known as *modern Physics* such as *quantum electrodynamics* or existence of extra dimensions emanated from String theory are among prime examples of such misconceptions. In fact, most of the theoretical laws within the framework of today's so-called modern Physics are illusions stemming from either wrong original formulation of perceived problems or forced-materialization of mathematical solutions in the real world. The irrationality in some of the proposed solutions of modern Physics, that have primarily been well-received as the most genuine contemporary theories, has even been openly admitted by physicists such as Planck, Feynman, and Heisenberg who pioneered the creation of the very same modern Physics.

Unfortunately, instead of being about understanding Universe, contemporary Physics has become making up arbitrary rules and formulae that partly fit different data. A clear example of such a law is known as Schrödinger wave function which is exclusively aimed

35

at atomic and subatomic scale structures. Schrödinger's equation can for most part be directed towards any desired solution by selective implementation of certain boundary conditions and pre-assumptions especially when seeking solutions in complex atomic systems. In fact, leaving few cases of the classical Physics aside, the established formulae in today's Physics are entirely based on empirical observations that are exclusively calibrated through extensive trial and error mechanisms.

STUMBLING JUSTIFICATIONS

Physics' state-of-the-art experimental equipment and its adoption of fancy terminologies such as Graviton, Magnon, Skyrmions, etcetera help indoctrinate the notion that it has already fully solved the old scientific problems and that it is rationally looking for answers to the emerging unknowns of Universe. Nonetheless, focusing on the fundamentals of modern Physics reveals that it has yet to answer many of the most basic questions regarding its devised principles during its relatively short history. A selection of such ignored questions that expose inherent shortcomings associated with Physics' offered justifications are highlighted here.

1. Why, as it was first reported in late eighteenth century, does a positively charged metallic sphere discharge when moderately heated while a negatively charged metallic sphere keeps its charge under the same conditions? Charge loss in the positively charged object occurs even when it is kept under vacuum isolation.

2. Through irradiation heating of an electrically neutral metal in vacuum, its electrons are emitted in what is known as *thermionic effect*. Considering the finite number of free

electrons inside such a metal, why doesn't electron emission stop or degrade after heating up the metal in vacuum for a long time? In a discussion also related to the previous question, after it is cooled down, such a metal is expected to have a net positive charge proportional to the number of electrons it has lost. The question is why after it is cooled down the metal would be electrically neutral?

3. Despite having their outer orbital shells fully filled by electrons, why do massive inert gases such as Argon, Krypton, Xenon, or Radon not only form covalent bonds with other atoms but also show relatively high valence numbers?

4. Why cannot He+1 or He+2 ions, under ambient conditions, make covalent bonds with other atoms and reach stability after filling their orbitals?

5. Based on Physics' principles, it seems logical that radical hydrogen atom would obtain the inert gas structure by acquiring one electron and turning into H-1. However, why is there no H-1 in ambient conditions and yet, on the contrary, the positive nucleolus tends to repel the only electron from its orbit and turn the atom into H+1?

6. N2 is a very stable molecule since its two nitrogen atoms have made three bonds. However, based on the same analogy, why despite good overlap of sp3 hybrid orbitals between two adjacent carbon atoms, doesn't C2 molecule exist?

7. Why is dissociation energy of every second pi bond in diatomic molecules, for example between carbon atoms in Acetylene, less than that of the first pi bond, despite involvement of similar hybrid orbitals in both bonds?

8. Since reflectivity in metals has been attributed to the interaction of its free electrons with light's electromagnetic field, it seems conceivable to manipulate the color of every conductor by adjusting the magnitude or direction of an applied electric field. Nevertheless, why wouldn't electrical charging of materials alter their color states?

9. After a photon hits the surface of a reflective metal, how does the spherical atom or valence electron know the incidence angle to accordingly give an exact reflection angle to the photon? More specifically, in surface analysis techniques of materials such as infrared spectroscopy, how does the energetically shifted photon come to know the reflection direction, especially after undergoing an inelastic energy exchange with electrons of the target material?

10. The intensity of a laser beam might be amplified after it enters a *gain medium* through what is known as the *stimulated emission* of the pumped electrons. How do the excited electrons in the amplifying medium know towards which direction they should radiate their energy once they recoil to lower states? Similarly, how in a *potassium dihydrogen phosphate* crystal, better known as KDP crystal, is the *second-harmonic* output beam created primarily in the direction of traversing laser pulse? This occurs while it is far more logical to have an isotropic emission of generated higher harmonic frequency by KDP after excited electrons are recoiled to lower energy levels.

11. Why does a laser beam diverge more significantly when for the same intensity the beam diameter becomes smaller, even though photons are considered to be bosons and should principally be neutral towards one another?

12. Why aren't diffraction patterns formed when laser beams of similar frequency and intensity but generated using separate setups are shone on each of the slits in Young's interferometer? If it is the wave nature of the light that leads to the fringe formation, regardless of the light source used, bright and dark fringes must be formed after the so called constructive and destructive superimposition of waves propagating from the two slits.

13. In Young experiment, interfering bright and dark fringes are observed when the dimensions of the slits are comparable with the wavelength of the incident visible light. Why is not the same experimental concept applicable to long radio waves entering slits with widths adapted to their wavelengths?

14. Why is magnetic field around a beam of electrons flowing in vacuum immeasurable, as opposed to the measurable magnetic field around a wire conducting an identical electron flow?

15. If it is the intrinsic spin of electron that creates its magnetic dipole moment, why is it not possible to produce a magnetic field out of spinning a macro-scale charged sphere around its axis?

16. If magnetism is indeed due to the electron-ordering patterns inside individual atoms, then why is there no single atom that can magnetically be absorbed by a magnet?

17. When electrically charged, a metallic sphere creates a single isotropic electric field. Why is it not possible to make an analogous magnetic sphere with only one single pole?

18. If it is the spin orientation of electrons that makes *oxygen* magnetic in its crystalline phase, why doesn't a similar magnetization occur to *fluorine* or *sulfur* whose atomic structures, compared to oxygen's orbital structure, have one extra electron in P-orbital and the same configuration of electrons, respectively?

19. If ferromagnetic, paramagnetic, or diamagnetic behavior of an element is solely related to the number of electrons in its outermost atomic orbital, why do some of the elements in Mendeleev table exhibit such a magnetic property while other members of the same group do not necessarily show the same property?

20. Why are not particles more massive than alpha emitted by radioactive atoms?

21. Why is it not possible to use a similar *gamma* source for studying different target nuclei in Mossbauer spectroscopy? For example, why is it not possible to use a 57Co source, which is very effective in studying 57Fe atoms, in studying 191Ir atoms? In theory, one could increase *cobalt* source's range of effective energy sweep by adjusting the oscillator's speed and thus use the same gamma source in studying nuclear energy levels of 191Ir atoms.

As religion, Physics will never admit its inability in answering any of these questions as it considers such admissions to be detrimental to its establishment. Yet, physicists will skillfully find ways to justify these and any other flawed aspects of their ideology by aptly switching between classical and modern concepts of Physics or by obscuring the issues using mathematical twists. At the end, nevertheless, no matter how much Physics' presented justifications are cherished by its constituent members, these justifications will remain meaningless to rational critics outside the faith of Physics.

CHAPTER THREE: MEASURISM

MATERIAL PROPERTIES

Matter and the variety of its properties are what making Universe. Materials in Universe are distinguishable through their comparative properties which are always referenced to at least a known point within the spectrum of the same property. Attributing a property such as *thick*, for example, to a material is only meaningful when a comparison is made with respect to a known *thin* material.

Due to their comparative character, all properties in Universe exist in binaries for which either side stands only to disambiguate the other side. Each property has a side which is unrestricted in magnitude while its other side indicates the lack of the property. For instance, in complementary binary properties such as *bright and dark* or *hot and cold* there are no limitations on the magnitudes of *bright* or *hot* properties as there can always be a brighter or a hotter property, while on the other side, *dark* or *cold* properties only refer to the absence of light or heat, respectively.

Complex properties in materials may appear after the combination of more basic materials. A good example of such complex materials is living matter which may be defined as any independent combination of basic materials that exploits the environment to its benefit and that actively tries to protect itself against outside threats. According to the unique aspects of each living matter, some of its properties might be unique to itself and inapplicable to other living matter.

Unlike the case of properties belonging to more basic materials, some complementary properties in living matter at first might seem to have unlimited boundaries on both sides. To name a few, such complementary binary properties may include *happy and sad, pleasant and unpleasant, patient and impatient*, or *good and bad*. However, either of the sides within any of these binary properties is an independent biological perception stemmed from a combination of living matter's structure and surroundings. Therefore, either of the sides within these presumably complementary binaries is an independent property which should be referenced to its own absence. For instance, sadness is not caused by the lack of happiness or vice versa. Sadness and happiness are independent properties which must be referenced to their own absence.

In mathematical terms, every property whether basic or complex, lies linearly within its infinitely extendable value on one side and its hypothetical absence on the other side. It is important to note, just like the scattered character of matter in Universe, all basic and complex properties take discrete values when observed or measured over their distribution spectrum.

LOCATION AND TIME

Location of matter introduces a different type of property in Universe. Complementary binary properties such as *east and west* or *above and under* lie in the category of location property. Similar to the case of basic or complex matter properties, properties referring to location are meaningful only when a comparison is made to at least one known reference location. However, unlike the earlier properties, neither side of the complementary binaries within the location property compares with the absence of the property. Instead either side of the complementary binaries belonging to location property is extendable to infinity. Every *beneath* might be *above* and every *front* might be *behind* according to assumed reference locations. This further implies that there are no limits to the boundaries of Universe.

Apart from the location property, *time* is the only other property of matter that is not measured against its absence and has unrestricted sides. In other words, neither side in complementary *past and future* binary property can be traced to a start or an end. More specifically, there has been no start to Universe and there will be no end to it.

The unlimited character of *location* or *time* binary properties further applies to their derivatives such as speed and acceleration. That means either side within *fast and slow* binary property might be stationary, slow, or fast depending on the speed and direction of an agreed reference coordinate system.

Whether properties of matter are extendable on one or both sides, the common aspect of all properties lies in their measurability. All properties are measurable relative to comparisons made with agreed reference points within the same property and there exists no property that is unmeasurable or returns a fixed value upon measurement in different reference systems. Likewise, there exists no matter whose properties are unmeasurable or fixed.

THE FOURTH DIMENSION

Even though, three dimensions are enough to locate matter within this world, to locate the world itself within Universe an extra dimension is needed. A fourth dimension within location property implies that despite human's experience of dealing with a unique world, there are unlimited worlds that are making Universe. Just like the discrete distribution of matter in the three-dimensional world, matter across the fourth dimension should exist in discrete patterns.

The fourth dimension of location is necessary to justify the fact that there are no limits on the largest and smallest making components of Universe, namely galaxies and atoms. The galaxies in the sky are not the biggest elements of Universe but are atoms creating the *world ahead* whose sky galaxies are themselves atoms creating another world ahead. Likewise, atoms which are making components of a world are galaxies in the sky of its *world aback*. In neither direction, the succession of ever repeating worlds has an end.

It is worth emphasizing that Universe has always been in and will always remain in equilibrium and that events in different worlds do not disturb its overall balance. Also, the fourth dimension implicitly implies that in Universe there exists no absolute vacuum void of matter and therefore void of properties.

Experience of the *time* concept in all individual worlds remains the same since time is just a terminological convention which accounts for a comparative rate of change in every world. However, when the rate of change is referenced to a given world, the relative rate of change for worlds ahead or aback would be drastically different.

Regardless of the rate by which events unfold in galactic or atomic scales, a full correspondence exists between components and their behaviors in both scales. The linear behavior of properties helps figuring out a crude correlation between rates of change happening in every two adjacent worlds. Assuming, as suggested by today's Physics, diameters of 10^{20} and 10^{-10} meters for a spiral-galaxy and a hydrogen atom, respectively, a rough size difference of 10^{30} between similar components of two successive worlds is calculated. Therefore, the rate of change in any world aback could safely be assumed to take place 10^{30} times faster compared to its adjacent world ahead. In a related analogy, using a spiral-galaxy's full rotation period of around 10^{12} seconds, the linear velocity of compounds on the periphery of a hydrogen atom are calculated to be approximately 3×10^8 meters per second which roughly equals the speed of electromagnetic radiation as it is promoted by today's Physics.

Universe is based on simple rules of materials and their properties. No material or property exists that returns a fixed value upon measurements with respect to different reference systems. In general, matter with an unmeasurable property is nonexistent. Therefore, the concept of having a unique-speed property for whatever entity, as observed from various coordinated systems, is absurd. Speed of light or any other moving matter is measurable and, contingent on the reference coordinate system used, may take all the values from positive infinity to negative infinity.

GRAVITY

The measurable character of properties in Universe invalidates the exception of a merely pulling force or a force whose magnitude is independent of assumed reference coordinate systems. Therefore, besides its pulling character, gravity should have pushing and dragging aspects. Although, pulling character of gravity is the main force within the galaxies, these are dragging and pushing characters of gravity which keep Universe from coalescing into a single accretion of matter. A balanced and ever-lasting Universe must contain comparable magnitudes of opposing forces.

Gravitational properties of worlds aback appear to be the main sources of all known forces in every world. In fact, all forces as known in today's Physics could satisfactorily be explained through certain gravitational interactions happening in atomic and sub-atomic scales. However, it must be emphasized that gravity force in every world is as much about the dynamics of gravity in the worlds

46

ahead as it is about gravitational dynamics of the worlds aback. In general, the parameters governing the gravity force within every world are solely decided by the interactions happening in other worlds. Therefore, any speculation on manipulating the gravity force or field lines, referred in today's Physics as *anti-gravity*, is utmost nonsense.

Considering mass dynamics, gravity force may exert anisotropically on different points of a shell encapsulating the mass. In other words, in addition to its known pulling force, the gravity force emanating from any moving mass drags the nearby matter proportionally in the direction of its movement. Applying the dragging character of gravity through including a mathematical term to the *universal law of gravity* will successfully explain the unknown behaviors of celestial entities observed within Solar system or outer space.

In the case of Solar system, Sun anisotropically drags the planets along the direction of its spinning plane. The closer the planets are to Sun, the stronger dragging effect of Sun they experience. This explains the *perihelion precession anomaly* observed in Mercury's trajectory as closest planet to Sun. The dragging force of gravity exerted by Sun on Mercury slightly accelerates the planet on its trajectory and leads to a more forward-occurring Perihelion along the spinning direction of Sun, after each period of the planet. The anomalous precession changes observed for Mercury cannot be satisfactorily explained using the universal law of gravitation or even utilizing the assumptions of Relativity theory.

A further indication of the dragging character of gravity within Solar system is observed in its planar configuration. The dragging force of gravity is strongest on the spinning plane of Sun and decreases to zero for objects located on either side of Sun's spinning axis. This is the reason, why planets within Solar system tend to rotate around Sun on its spinning plane.

The dragging gravity also successfully explains the anomaly seen in the angular velocity of stars within the discs of spiral-galaxies. In this anomaly, the observed speeds of stars located towards the outer edges of the discs in spiral-galaxies do not follow the universal law of gravitation. In fact, these stars rotate much faster than what is calculated from the gravitation law. This anomaly has been the main reason behind adoption of the superstitious concept of *dark matter* and its entire descendant theories. The concept of dragging gravity justifies this anomaly by stating that each star is dragged along by its neighboring stars as they rotate around the disc. Within the outer disc, the dragging contribution of nearby stars eventually leads to a leveling of angular velocities among all stars forming the disc.

Furthermore, by including the dragging force of gravity in the universal law of gravity, other anomalies such as the ones observed in the rotational behavior of satellites around planet Earth, known as *flyby* anomalies, could be explained.

Laboratory Evidence

Over time, the developments in characterization and observational techniques and equipment will make it more probable to find the right answers to an increasing number of different phenomena in Universe. Nonetheless, even using available technologies, there are few experiments that, if performed, reveal true answers to quite a few misunderstood or misinterpreted phenomena. Inevitably, such experiments will stand to invalidate several fundamental principles of Physics.

One such an experiment would be concerned with revealing different velocities of photons forming diverse regions of the radiation spectrum. Measuring speeds of light emitted from monochromatic sources, for instance lasers, will reveal that blue-color light travels slightly faster than green-color light while green-color light travels slightly faster than red-color light. In fact, these are different speeds of light colors which make these photons interact with matter distinctively, and not their assumed frequencies or wavelengths as claimed by Physics.

Another revealing experiment would be resolving and distinguishing different geometric profiles of an atom. In such an experiment, low-temperature scanning probe microscopy setups can resolve face-on or edge-on profiles of atoms within certain crystals, especially under the optimal application of external magnetic or electric fields.

THE ANISOTROPY

The opposing sides of properties seem to suggest that Universe moves towards a static equilibrium far from entropy. Since on the contrary, Universe is already at a dynamic equilibrium, Universe's governing rules should at least contain a single anisotropic character. Such an anisotropic behavior is also needed to explain some of the selective trends observed in Universe such as the structural asymmetry between the mates of living matter or bending of negatively and positively charged particles towards opposite directions while traversing through an applied magnetic field. Such anisotropy may further explain why mass in spiral-galaxies or celestial objects tend to occur around a similar value.

The only anisotropy in Universe appears to be caused by the matter being absorbed to or repelled from galaxies and celestial objects in selective directions. Specifically, clockwise rotating galaxies once looked at face-on in an absolute majority of cases lead to the creation of a repelling gravitational force against the observer away from the disc and a simultaneous pulling gravitational force on the other side towards the disc. Negatively and positively charged galaxies or celestial objects form opposite directions of pull-push anisotropy which explains their attraction towards opposite polarities. It is probably safe to assume that the ratio between galaxies obeying and disobeying this general rule is comparable with the ratio between the occurrence of right-handed and left-handed individuals.

LIFE

From a personal perspective, all properties might be referenced to their absence, including position and time. This is because from an individual point of view, every person can be assumed at the center of Universe in which every object is considered as near or far. The past memories and future aims are making hem what the person is while each expected event may occur to hem within a soon or a late timespan. No one is able to remember hes birth and no one will ever remember hes death. Everyone lives in the middle of hes past and future.

Someone has written these words that I can continue or stop reading. These are my thoughts that make Universe in which all the events are arranged to support my existence. I am always where it is called *now* and I am unable to end it the same way I could have never started it.

ANNEX: UNIVERSE

Much of the current human knowledge about Universe concerns atomic scale interactions. Even though, most of the effects and their causes within atomic scales are well observed and documented, the justifications presented for these interactions are overwhelmingly based on flawed perceptions. On the other hand, when it comes to human knowledge of Universe in galactic scales, even the simplest events and their causes are unknown let alone the existence of proper explanations for them. The main reason behind the naïve human knowledge of the galactic-scale phenomena lies in their extremely-slow pace of development relative to rates of change experienced by human.

Human's immature knowledge of galactic phenomena involves even the most basic properties of celestial entities including their masses, sizes, or distances. While Physics employs various methods to estimate mass, size, or distance of celestial entities, all these methods rely on highly questionable assumptions in their estimations of these properties. For instance, using luminosity in determining distance of a galaxy implies that all galaxies with similar masses are already presumed to illuminate with the same brightness. Another mistakenly used approach employs Hubble's law that assesses the distance of galaxies from Earth by relying on the imaginary Big Bang theory and its Expansion-theory offspring. Hubble's law presumes that the distance to a galaxy and galaxy's recessional velocity, as seen from Earth, are proportional. Therefore, based on Hubble's law, the larger the red-shift observed in the emission spectrum of a celestial entity, the farther it is from Earth.

Knowing all worlds in Universe replicate similar experiences, the most effective way to understand a world and its governing rules would be to make a one-to-one correspondence between its atomic and galactic scale components. Thereby, to find out the unknown issues in either of the atomic or galactic scales, the known knowledge from the corresponding match would be equally applicable. Although, the limited knowledge of atomic and especially galactic scale phenomena makes it difficult to confidently assign every single of these one-to-one analogies, such an approach would significantly improve human understanding of Universe. Therefore, in the following sections, the most appropriate of such correspondences in accordance with the available knowledge of Universe are made in a rather simplistic way. These correspondences are further used to mainly explain much of the fundamental interactions in atomic scales and to a lesser extent to apply the existing knowledge in atomic scales to galactic properties and effects. While during the arrangement of the contents in the following sections it has been tried to sequentially prepare the ground for every upcoming subject, it should be borne in mind that most topics contain overlapping concepts.

HYDROGEN

Considering the overwhelming abundance of both spiral-galaxies and hydrogen atoms in the world, hydrogen atoms best correspond to spiral-galaxies. However, unlike the current perception that mass and size properties among spiral-galaxies differ significantly, in fact, either of these values may only slightly vary from an ideal equilibrium. The equilibrium state in a spiral-galaxy is achieved when it has no or least desire in losing or gaining celestial bodies.

Based on various experimental results such as consistent bending of trajectories of ionized hydrogen atoms, while passing through electric or magnetic fields, it should be the mass values of hydrogen atoms that occur around an equilibrium value. Mass values in isolated hydrogen atoms are more likely to be closer to the equilibrium value than to occur with large deviations. Since the dimensions in each atom are closely entangled with its mass, the size of hydrogen atoms should also occur around an average value with little deviations.

HEAT

Heat in every world is the interpretation of an increased state of vibration or movement of celestial entities in the immediate world aback. These vibrations or movements are the result of gravitational perturbations caused by external matter increasing in amount within the inter-celestial medium. External celestial entities that may cause the perturbations range from simple celestial bodies such as dust, planets, and stars to celestial objects such as galaxies and other groupings of celestial bodies. The gravitational perturbations disturb the otherwise largely smooth rotation of galactic discs around the bulges. In practice, by heating up a substance, external matter is injected into the vicinity or halo of atoms perturbing the existing gravitational harmony among components of each galaxy and therefore among neighboring galaxies. Hence, an increase in temperature is an indication of faster, and often less rhythmic, vibrations of galaxies and their constituent components compared to their equilibrium state.

On the contrary, cooling down a substance means withdrawing the excess intergalactic matter from the halo and vicinity of galaxies in their immediate world aback. Galaxies residing in an environment with minimum external celestial entities, experience least gravitational perturbations. These galaxies may maintain the direction of their disc rotations or precession axes for long durations of time, according to the equilibrium they have established with neighboring galaxies.

NEUTRONS

Because of the strong penetrability of a *neutron* particle through most of materials, its lack of interest in sensible response to magnetic or electric stimuli, and its emission by lightweight atoms once bombarded by speedy ionized hydrogens, it is safe to assume neutrons as hydrogens which are gravitationally perturbed beyond a reversible state. Considering the available knowledge about celestial objects, currently a neutron particle best corresponds to an elliptical dwarf galaxy.

Hydrogen's family of isotopes, namely hydrogen itself as well as deuterium and tritium, provide the most direct evidence to the existence of neutron particles. In the absence of such evidence it might have been argued that neutrons were just provisional transformations of hydrogen atoms after being ejected from target atoms by impinging ions. Deuterium and tritium isotopes are grouped with hydrogen since their chemical properties mostly resemble one another as it is seen in heavy and tritiated water formations, respectively. Concerning the equilibrium configuration of constituent components, it is probably best to assume that in tritium the nuclear components are lined up as neutron-hydrogen-neutron with the two neutrons making contact to hydrogen on the opposite sides of the hydrogen disc while the mass centers of all three components lie on the same line. In the case of deuterium, one of the neutrons is absent and therefore the isotope is gravitationally less perturbed in comparison with tritium, leading to a more stable isotope.

Free neutron particles have been mainly detected after external stimulation of lightweight atoms through their bombardment by proton and alpha particles. Considering the unknown effects of ejection mechanisms and subsequent behaviors of ejected neutron particles, there is insufficient evidence to draw parallels between the levels of instability caused by the stray neutrons and constituent neutrons residing within atoms. The arrival of external neutrons, especially the ones with optimized kinetic energies, has been observed to have large destabilizing effects on various isotopes while the residing neutrons not only might have little or no destabilizing effect, but might even contribute to the stabilization of some isotopes.

Although, highly unstable hydrogen passed beyond the reversibility point is presented as the best model for the structure of a neutron, the experimental observation of magnetic field effect on neutron alignment indicates at least a low degree of harmonic behavior among the celestial bodies forming neutron particles.

ATOMS

Atoms more massive than hydrogen correspond to galaxies more massive than spiral-galaxies. The simplest case of a union in galactic scales occurs between two closely placed celestial objects when under negligible gravitational disturbances from the surroundings. In such a situation, the two celestial objects find an equilibrium state in which both seem to lose a small fraction of their celestial bodies. In the case of two spiral-galaxies being in such a state, they form a structure which is known as hydrogen-hydrogen binary while in the case of a spiral-galaxy and a dwarf galaxy it leads to the formation of a hydrogen-neutron binary structure. It seems feasible that under certain conditions, even ternary or quaternary constituent elements may combine and create more massive structures, though this is much less likely because of further complications caused by the extra gravitational perturbations involved with the higher number of constituent entities.

In the absence of strong gravitational disturbances, these binary structures stay intact. However, in the presence of excessive celestial bodies, based on how stable these structures are, some of them may end up acquiring an encompassing disc of celestial bodies. The gravitational interaction between the core components and encompassing disc leads to the formation of a far more stable compound, compared to what would have been achievable in the absence of encompassing disc. Such combinations of core components and their encompassing discs make what are known as atoms.

Helium atom makes the least complicated example of atoms after hydrogen. The best way of visualizing the dominant isotope of helium atom seems to be through a combination of two spiral-galaxies and two dwarf galaxies as core components that are all confined by a joint disc of encompassing celestial bodies. Although, the mass of encompassing disc appears to be small, compared to core components' mass forming the bulge, it plays a significant part in the overall stability of helium.

According to the current knowledge of atoms, two conjoined spiral-galaxies form a stable pair known as hydrogen-hydrogen molecule. Therefore, neither of the spiral-galaxies forming the pair within helium core tends to form a bond with external spiral-galaxies. This explains why under ambient conditions helium strongly disfavors forming a chemical bond with other atoms and thus is known as one of the *noble gases*.

Neon makes another noble-gas configuration example whose most stable isotope structure is created when ten spiral-galaxies and ten dwarf galaxies are gravitationally confined within a joint encompassing disc.

Just like the fact that values of mass in all atoms are roughly a natural-number multiple of the mass of single hydrogen, mass of galaxies should be an approximate natural-number multiplication of the mass of single spiral-galaxy in equilibrium too. However, the estimated mass of atoms deviates from an exact natural-number multiplication of a spiral-galaxy due to specific bulge dynamics and especially due to varying disc-masses.

The total number of spiral and dwarf galaxies forming the bulge of atoms is an important factor determining how stable an atom can be. Some specific numbers, known as *magic numbers*, usually result in symmetric distribution of bulge components and therefore increase the overall stability of the corresponding isotopes.

It is worth emphasizing that since Universe is in equilibrium, the creation of massive atoms from smaller ones should not change its balance towards a Universe with more abundant massive atoms. There are by far more spiral-galaxies formed from celestial bodies than massive atoms formed from fusing spiral-galaxies. The overall outcome is that the relative fractions of atoms in the world always stay around their all-time equilibrium ratios.

Spiral and dwarf galaxies within the bulges of atoms seem to be constantly moving. Thus, the backscattering of accelerated charged particles from the bulge is not necessarily an electrostatic encounter but could also be a gravitational repulsion of the incoming particles. Accordingly, under certain encounter conditions, scattering always takes place whether impinging particles are charged or neutral. Respective experimental examples of these situations are observed in Rutherford and neutron backscattering experiments.

In practice, the constant coverage of the bulge components by the sweeping disc plane leads to the formation of an effective galactic sphere around atoms. Only those celestial objects which constitute a small fraction of the host galaxy's mass can enter its galactic sphere. Arrival of more massive celestial objects would cause a mutual repulsion should they approach the galactic sphere. Such repulsions lead to temporary destabilization of both entities and thus loss of some fractions of their celestial bodies depending on the nature of the encounter.

Besides the gravitational confinement created by a rotating disc, the dynamics of the core components within the bulge seem be equally responsible for the overall stability of atoms. In some situations where the rotation of disc plane around the bulge is hampered by interlocking with neighboring atoms, constituent components of the bulge move faster within the bulge which is a consequence of the preservation of rotational momentum in atoms. The overall movements of components within the bulge follow patterns which are decided by atom's unique configuration and the nature of its interactions with neighboring celestial entities. Usually, the internal dynamics of bulge and disc cannot disrupt the gravitational balance of an atom. However, in atoms with unstable bulge configurations, the gravitational confinement of the encompassing disc may fail in properly retaining the components of the bulge. Thus, as it is seen in unstable isotopes and most massive atoms, bulge constituent components occasionally manage to escape the bulge in materials which are referred to as *radioactive*. Such escapes usually decrease the instability in these atoms by reducing the overall gravitational repulsions within the bulge. Apart from dynamics of the bulge and its encompassing disc, the conditions dictated by adjacent celestial entities also affect particle emission patterns in radioactive atoms.

The ejection of hydrogen and helium from radioactive atoms, known respectively as proton and alpha particle emissions, is an indication that these two atomic configurations are the main building blocks forming the bulge of more massive atoms.

The evidence of massive atoms' nuclear structure will sooner or later be established through the development of increasingly more sophisticated astronomic observational-tools which optically resolve the bulge components within big galaxies. However, should the Milky Way galaxy be a hydrogen atom located in an isolated constellation of hydrogen gas faraway from massive atoms, human striving to find conclusive evidence on the presence of massive atoms will become even more challenging.

BONDS

Every two closely located spiral-galaxies may form a joint gravitational equilibrium known as a *covalent bond*. In the case of two isolated spiral-galaxies this leads to the formation of H2 molecule. Gravitational perturbations increase the chances of bond formation among neighboring atoms by moving and rotating the spiral-galaxies to the right position and angle.

Just like the bond formation between isolated spiral-galaxies, spiral-galaxies located inside neighboring bulges of massive atoms may form a covalent bond across the two atoms. In a single bond formation between spiral-galaxies located inside the bulges of neighboring atoms, the distance between the atoms on the line connecting the two bond-forming spiral-galaxies would be at its minimum. The first bond formed between a pair of spiral-galaxies across neighboring bulges is known as *sigma bond*. The line of bond connecting the two bulges occurs away from disc planes belonging to the engaged atoms and does not cut through or over them.

Creation of the second bond between another pair of spiral-galaxies across the already-sigma-bonded bulges leads to the formation of two equally distanced bonds across the bulges. Obviously, in the case of having double bonds between two neighboring bulges, the competition between the two pulling bonds, located across the galactic hemispheres facing the other atom, results in the formation of lengthier bonds compared to the situation where only one bond existed. This explains why the second bond, regarded as *pi bond*, has been wrongly perceived to result in a weaker bond compared to the sigma bond. The formation of the second bond leads to an elongation of the first bond as the two bonds pull atoms across the two bulge hemispheres towards each other. The spatial position of either bond within the bulge at any given moment mainly depends on the restrictions imposed by the gravitational interactions among constituent components of every bulge and therefore slightly varies around an average value. A noticeable consequence of second bond formation between two neighboring atoms would be locking of the engaged bulges on a planar surface containing the two bond lines.

Formation of third bond between the two neighboring bulges leads to an elongation of the existing two bonds to a new equilibrium length which is also shared by the third bond. These three equally-distanced bonds are almost symmetrically located on the corners of an equilateral triangle around the position of minimum distance between the two bulge spheres. The exact arrangement of the bonds depends on the configuration and dynamics of the adjoined atoms including the locations of their disc planes.

A longer distance between spiral-galaxy pairs forming the bonds, leads to weaker individual bonds compared to the situation where less bonds existed. In other words, the *dissociation energy* needed to break the first of individually paired spiral-galaxies decreases when the number of bonds between two neighboring atoms increases. Thus, the dissociation energy for remaining bond or bonds increases after breaking each further bond among the two bonded atoms.

Carbon atom provides a good example of bond formation mechanism. Carbon's atomic structure places it somewhere between those of helium and neon and its stable isotope contains a total of twelve spiral and dwarf galaxies at its bulge. In its equilibrium, four of the six spiral-galaxies may geometrically occupy corners of a tetrahedron within the bulge. The remaining two spiral-galaxies occupy the center location of tetrahedron as a bonded pair. It seems two out of the four spiral-galaxies located at the corners of the tetrahedron can form bonds among themselves or with the outside spiral-galaxies. Therefore, a carbon atom may make two or four external bonds with external atoms.

In the case of two carbon atoms forming chemical bonds together, the maximum number of formed bonds would be limited to three since one of the spiral-galaxies in both atoms will always be located on the opposite side of the bulge spheres facing away from the bonded carbon atom. This is the only reason why there are no carbon-carbon molecules with four bonds, something which cannot be explained by current Physics' concept of *orbital hybridization* in carbon atoms. Similarly, for all relatively small size atoms the maximum number of established bonds among every two similar atoms remains three.

The maximum number of bonds an atom may form with surrounding atoms is known as *valence number*. In small size atoms, direct correlations exist between the valence numbers and the number of constituent spiral-galaxies within the bulges. The valence number of atoms neighboring carbon in periodic table decreases or increases, respectively, once a spiral-galaxy is removed from or added to carbon bulge. Besides the valence number, the spatial configuration of neighboring atoms or the distribution of spiral-galaxies within their bulges might also affect the maximum number of bonds made by an atom.

Large atoms generally show a higher valence number compared to smaller size atoms. This is because the rotating disc in large atoms is unable to as tightly confine the bulge components compared to its confinement of bulge components in small size atoms. At the same time, large atoms contain a higher number of spiral-galaxies within their bulge. This concept further explains why large atoms belonging even to the so-called *inert gases* group of elements may form chemical bonds with other atoms.

Various physical and chemical properties of matter depend on interactions arising from the dynamics of its constituent components. Mass of atoms, disc sizes, location of the encompassing disc with respect to other bulge components, angles under which spiral-galaxies form external bonds, frequencies by which internal or external bonds disconnect and reconnect, and whether in the latter case the new connections are made among the same or different spiral-galaxies are among the main reasons driving chemical and physical properties in various materials. In most media, as seen for example in quartz, even the macroscale crystal geometry has been observed to affect the harmonic oscillations of the residing atoms. The complications underlined by these factors explain why crystallography alone has been unable to answer many fascinating natural phenomena such as the occurrence of patterns conforming to Fibonacci sequence.

CHARGE

Excessive amounts of celestial bodies concentrated within a small space may agglomerate in the form of electron particles. Thermionic effect is an example of such a situation where electron particles are continuously formed from the agglomeration of excess celestial bodies present at the intergalactic medium which in this case are subsequently emitted. The excess celestial bodies might be injected as a result of heating or might be released by the intergalactic medium through stirring of its gravitational balance by, for instance, magnetic field induction. Thus, unlike the claim of today's Physics that number of electrons in a medium is limited since constituting atoms contain a certain number of electrons in their shell, in fact, there are no limits on the number electrons created by different media since each individual atom can generate new electrons for so long as it is supplied with external celestial bodies.

Based on the current knowledge of celestial objects, at the moment electrons in every world best correspond to globular-clusters in the immediate world aback. This is because globular-clusters beside the less massive open-clusters are the only celestial objects found within the halo of galaxies. However, considering their more massive structure compared to open-clusters, globular-clusters should be the ones acting as electrons in the world ahead. The fact that globular-clusters are mostly seen within the halo of galaxies also helps justify the *photoelectric effect* phenomenon in which photons with adequate kinetic energies are able to eject electrons.

Just like mass of other celestial objects, it is best to assume that mass of globular-clusters in their equilibrium tend to occur around an average value. A globular-cluster in its equilibrium-mass state seems to be neutral and may act as a negatively charged electron only when it is lacking mass to reach equilibrium. A mass deficient globular-cluster anisotropically pulls on outside matter to absorb celestial bodies into its structure and to compensate its lack of mass. The anisotropic behavior of charged globular-clusters necessitates the assumption of a disc-like structure that is pulling external celestial bodies from one side and repelling them from the other side, with respect to the direction of disc rotation. The anisotropic pull-push character of electron, defined relative to its disc-rotation direction, has experimentally been observed in *spin valves* where only electrons with a specific disc-rotation direction may pass a ferromagnetic medium. The anisotropic pull-push direction towards external mass in negatively chargeable celestial objects such as globular-clusters is generally opposite to that of the positively chargeable celestial objects such as spiral-galaxies or positrons.

Electron charge formation or loss may occur all over the halo of atoms and that is different from the charge concept when applied to isolated atoms. Charge condition in any isolated atom is decided by the individual spiral-galaxies residing within the bulge. This consideration is required to explain experimental observations of charge occurrence in ionized atoms like singly or doubly charged helium atoms. While a spiral-galaxy in its equilibrium state is always neutral, a mass deficient spiral-galaxy behaves as positively charged since it pulls on external matter in order to compensate its lack of mass and reach the equilibrium state. The magnitude of pulling force exerted by a positively charged spiral-galaxy seems to be unrelated to the level of its mass deficiency. This is probably because while a higher mass loss leads to a stronger gravitational desire for its compensation, at the same time it has already caused a less overall gravitational pull force since the gravitational pull is directly proportional to the overall mass dynamics of the spiral-galaxy. Obviously, the same principle applies to other chargeable celestial objects.

Globular-clusters within the halo of atoms might be ejected due to various gravitational perturbations. The mass deficiency occurred due to the lack of globular-clusters within the periphery of atoms results in a gravitational pull of nearby globular-clusters or stray celestial bodies. This gravitational desire that tends to compensate the lack of globular-clusters or celestial bodies within the halo of atoms is known as positive electrostatic charge or *holes*. Contrary to the deficiency of globular-clusters within the halo of atoms, their excess leads to a gravitational push of these particles by host atoms. The extent to which a material is intent to absorb or repel globular-clusters amounts for the strength of what is known as positive or negative potential, respectively. The charge-dependent desire of materials to absorb or repel stray celestial bodies has experimentally been employed to adjust temperature through modifying electrical currents and, the other way around, to create electric currents through modifying temperature. This mutual correlation between current and temperature states in materials is known as Peltier-Seebeck effect.

The concept of charge as explained here is best evidenced by the fact that moderate heating of positively charged metallic objects leads to their discharge as the excess celestial bodies introduced to the object compensate for its lack of mass or globular-clusters. On the other hand, in the case of negatively charged metallic objects, moderate heating does not affect their charge state.

CONDUCTION

Conductivity in materials seems to be a direct consequence of their constituent-atoms' ability to freely change the planes of their rotating disc around bulges in an environment of closely packed atoms. A material can conduct electrons in the direction of linearly-aligned neighboring atoms when the atomic discs are able to position themselves perpendicular to the direction of current flow and in such a way that all discs rotate towards the same direction. In conductors, the direction and strength of the gravitational direction among neighboring atoms remains similar and is decided by the kind and strength of the electric field applied to the media.

Hopping of electrons between neighboring atoms within a material appears to be hampered when disc planes are locked in nonparallel fashions or when their rotations occur towards opposite directions. Thus, resistance of a medium against the flow of electrons seems to mainly be an issue of the difficulty by which discs in neighboring atoms are able to reorient around their bulges. The bonds made among atoms in metallic substances appear to be easily relocatable around the bulges and therefore their positions does not hamper the reorientation of atomic discs. The ease by which bonds can be relocated around the bulge apparently increases with the growing size of the bulge. This is probably the main reason why more massive atoms tend to show higher metallic properties compared to atoms with small masses.

The reorientable character of disc planes in metallic materials also explains their ductile property. However, not every conductive material is necessarily ductile. A good example of a non-ductile conductor showing very high electron mobility is *graphene*. Graphene is a sheet of carbon atoms in which each constituent carbon atom forms three covalent bonds with three neighboring carbon atoms. In other words, only three out of the four available spiral-galaxies within the bulge of carbon atom form bonds with spiral-galaxies across the bulges of neighboring carbon atoms. The engagement of three spiral-galaxies in bond formation among neighboring carbon atoms is primarily evidenced by graphene's planar structure. In graphene, due to the partial freedom of disc rotations in carbon atoms, they can freely reorient up to the threshold required for the effective flow of globular-clusters. Interestingly, on the other hand, carbon atoms act as one of the best ever-known insulators when forming four covalent bonds with four of their neighboring carbon atoms. This configuration of carbon atoms leads to the formation of *diamond* whose good electrical insulation is due to the locking of disc planes in their positions as a direct consequence of four covalent bonds being made between each carbon atom and its neighboring carbon atoms.

MAGNETIC FIELD

Magnetic field line distributions in any medium corresponds to the trajectories traveled by celestial bodies after they are temporarily detached from discs of atoms while mainly maintaining their initial angular momenta. The gravitational perturbations leading to such detachments are primarily initiated by the passage of electrons through atoms. Since atoms can only anisotropically allow the passage of electrons, discs of atoms participating in the current conduction maintain a similar in-plane rotation direction. Therefore, celestial bodies expelled from atomic discs only rotate towards the direction as dictated by the direction of current flow.

It must be noted that an expulsion of celestial bodies from the discs of globular-clusters themselves during gravitational encounters, similar to that of atoms, cannot be ruled out. However, the magnitude of such magnetic fields would be negligible when compared to the known scales of magnetic fields generated by galactic discs. Therefore, contrary to the claim of Physics which associates the generation of magnetic field to the movement of charged particles, magnetic field generation is almost fully a property of the medium which arises in response to the transfer of electron current through it. In practice, no magnetic field is generated around a beam of electrons or any other charged particles traveling unhindered in vacuum. This also explains why, contrary to the claim of today's Physics, magnetic field cannot be generated by spinning charged-objects, no matter how much their charge levels or spinning speeds are.

As soon as the electron-flow through an atom is stopped, discs reabsorb a comparable amount of celestial bodies they had contributed to the generation of magnetic field. During the resettlement process of celestial bodies, the celestial bodies forming the magnetic field might be exchanged among discs whose rotation planes are approximately coplanar. Therefore, the final settlement of celestial objects might not reflect the original contribution of each atom to the magnetic field generation which in turn leads to the creation of new local equilibria.

The loss-gain mechanism of celestial bodies in galactic discs explains why no magnetic monopole exists and that there are always double poles associated with every magnetized system.

Components within coplanar discs of two neighboring atoms, in the absence of any surrounding gravitational disturbances, tend to rotate in opposite directions. That is because only then the dragging gravitational force where the two discs are tangentially closest would minimally affect the gravitational balance of the rotations in the two atoms. Extending this concept to the discs of atoms on any cross-section of a conductor during current flow would reveal that coplanar neighboring discs should revolve in opposite directions. In other words, for any cross section of the conductor carrying current, roughly speaking half of atoms revolve clockwise while the other half revolve counter-clockwise. The experimental demonstration of this effect is known as Ruderman-Kittel-Kasuya-Yosida effect in which addition of each layer of atoms is seen to reverse the magnetic field direction around the last atomic layer.

Despite their minute mass, the detached celestial bodies forming the magnetic field can influence the directions of much more massive celestial objects due to the strong adherence of these celestial bodies to the media generating the magnetic field. The evidence of such phenomena is known as Lorentz force in which the path of a traversing negatively or positively charged particle bends as magnetic field is applied to the particle under an angle. For instance, an electron entering a top-down applied magnetic field bends rightwards, as referenced to the traveling direction, while if it were to enter a bottom-up magnetic field it would bend leftwards. This occurs because when the globular-cluster enters the magnetic field, its entering-disc's front-edge gets aligned so that it rotates towards the same direction as the bombarding direction of celestial bodies forming the magnetic field. Thereafter, due to its anisotropic absorption character, globular-cluster pulls on celestial bodies on its right side while repelling them on the left side. A positively charged spiral-galaxy or positron under similar conditions would bend towards the opposite direction due to its converse gravitational anisotropy direction. Based on the same principle, in what is known as Lenz's law, application of a gradient magnetic field to a conductor results in an anisotropic transfer of electrons and thereby a forceful alignment of medium's atomic discs along the direction of current flow.

The influence of magnetic field on the resistance of certain conducting environments, in phenomena which are incompatible with justifications presented by Hall effect within today's Physics, provides another evidence on the role of disc plane reorientations on electron transfer. The alignment imposed on the direction of disc planes via application of certain magnitudes of external magnetic field has experimentally been demonstrated to improve the mobility of electrons in some materials such as graphene. The aforementioned effect of magnetic field is counterintuitive as normally any magnitude of applied magnetic field should lead to a decrease in mobility due to its effect on bending and scattering of electrons away from their traveling direction. The increase in the mobility of electron current is probably because of an increase in the number electrons that successfully pass any given cross section of the conductive medium due to the forceful alignment of a higher number of atomic discs after the application of external magnetic field, whose magnitude may create different interaction dynamics among neighboring atoms forming the conductor.

The assumption of magnetic field generation concept as such further explains paramagnetic, ferromagnetic and even peculiarly known diamagnetic behaviors as observed in certain materials. While disc planes within a paramagnetic medium need to constantly be under an externally applied magnetic field to maintain their alignment, in a ferromagnet they may maintain their alignment direction even after the removal of external magnetic field. Contrary to the characteristics of disc planes in paramagnets and ferromagnets, in diamagnetic media apparently the disc planes are interlocked among neighboring atoms in such a way that the direction of generated magnetic field from the material defies the force generated via the application of externally applied magnetic field. All these magnetic behaviors are consequences of disc engagement behaviors among nearby atoms forming the material. While paramagnetic behavior can usually be considered a consequence of individual atomic response, largely independent of neighboring atoms, the ferromagnetic and diamagnetic behaviors appear to be the results of collective reactions coming from residing atomic-colonies within the matter in response to externally applied magnetic fields.

With regards to superconductors, Physics claims there is no resistance to the current flow inside a superconducting material. All measurements leading to such a claim refer to the extremely stable magnetic fields created around the superconducting environments as proof of the non-diminishing current. However, the stable nature of magnetic fields generated by superconductors is more likely to be a consequence of interlocked atomic disc-planes that are unable to regain the celestial bodies they lost during magnetic field generation. Possibly, the interlocking of atomic disc planes occurs after they are cooled down below the superconducting temperature of the material while contributing to the magnetic field generation. The interlocking of atomic disc planes in superconducting material holds for so long as their alignments are not irreversibly disturbed by the injection of extra celestial bodies whether in the form of heat or external magnetic field.

LIGHT

Light particles best correspond to open-clusters mainly because they are the least massive celestial objects seen within the sweeping space of galactic disc planes around bulges. Open-clusters seem to be colonies of celestial bodies with fluid gravitational behaviors that, unlike chargeable celestial objects, demonstrate identical gravitational behaviors on both sides of their discs. In the event of gravitational perturbations occurring in their host galaxies, open-clusters are let loose of their gravitational connection and follow linear trajectories with velocities equal to those of their final angular speeds prior to the ejection from galactic spheres. Similar to every other existing entity in Universe, and unlike how it has been promoted by today's Physics, open-clusters have mass and it shouldn't come as a surprise that they are affected by gravity.

An open-cluster entering the sphere of an isolated galaxy is gravitationally forced into a radius for which it would have the right orbital angular speed based on its entering linear speed. Thus, a higher speed open-cluster arriving at the sphere of a galaxy revolves the bulge on a smaller radius compared to a lower speed open-cluster. This, for example explains why contrary to the propagation of various light colors in vacuum, it takes a ray of blue-color light longer times to pass through transparent media, compared to a ray of red-color light. While passing through transparent materials, compared to the relatively slow red-color open-clusters, blue-color open-clusters experience more drastic trajectory changes when entering the sphere of atoms. This is because blue-color open-clusters have to locate on orbits closer to the bulge and therefore often have to deviate more towards the bulge before subsequently leaving the atomic spheres. The larger deviation, inevitably results in longer overall travel distances of blue-color open-clusters compared to red-color open-clusters in various transparent materials. Considering the small difference between the speeds of red-color and blue-color open-clusters, the longer travel distances through any medium of transparent atoms implies a longer travelling time for the faster blue-color open-clusters than red-color open-clusters.

Encounters between a traveling open-cluster and a galactic disc may follow one of the following two trends. Should the rotating disc of the galaxy at open-cluster's traveling radius be free of celestial bodies, the open-cluster would pass through the plane of rotating disc. Such a galaxy would be considered *transparent* for all traversing open-clusters of the same speed. On the contrary, should the rotating disc of galaxy at open-cluster's traveling radius contain a dense layer of celestial bodies, the entering open-cluster collides with the disc material and dissipates into the disc. Such a galaxy would act as *opaque* for all incoming open-clusters of a similar speed. In practice, however, the passage or blockage of an open-cluster during an encounter with host atom's rotating disc is to some extent a probabilistic process. The probabilistic nature of such encounter is due to the fact that besides the density of celestial bodies present at the position of encounter, the relative angle under which the open-cluster has entered the galactic sphere also plays an important role in determining the fate of incoming open-cluster. It is possible that an open-cluster enters a galactic sphere on a trajectory that is parallel to the disc plane and therefore misses any collision with the disc. In such a situation even if the atom acts as opaque for the traversing open-cluster, it cannot block the passage of open-cluster. It is obvious that in this case, increasing the number of atoms on the way of travelling open-cluster increases the chances of its absorption by the target atoms. On the other hand, any open-cluster encountering a target atom with a trajectory covering a line on the same plane as that of the atomic disc's, is absorbed by the target atom after undergoing an edge-on collision, regardless of atom's transparent or opaque character while receiving such open-clusters under other angles.

By assuming an uneven spreading of celestial bodies across different radii of the disc, the transparency and opacity of target atoms can plausibly be explained when receiving incoming open-clusters. The thickness pattern of the celestial bodies spreading across the disc, to a good approximation, can be assumed to be a function of radii as measured with respect to galaxy's center. The variations in the density of celestial-bodies might be both due to internal dynamics of the galaxy or due to its interactions with neighboring galaxies. Although concrete instances of uneven spreading of celestial bodies on concentric or near-concentric radii in galactic scales have yet to be established, planetary systems like Uranus and Saturn clearly show such distributions of matter via their possession of discrete, concentric, and coplanar rings.

It is worth emphasizing that orbits containing uniform thicknesses within the atoms do not necessarily form perfect circular or elliptical patterns. In fact, based on the specific mass distribution and dynamics of the galaxy's bulge or its gravitational interactions with the surrounding celestial entities, the thickness patterns within a galaxy's disc might be more complex.

Interactions of radio frequency, visible, and X-ray bands of radiation with matter constitute three unique examples of open-cluster behaviors once interacting with atoms. Radio frequency, or RF, is referred to the wide frequency band which is nowadays employed for numerous telecommunication purposes. After entering the galactic sphere of target atoms, open-clusters forming the RF band would gravitationally be located on the very outer edges of galactic spheres. Since the outermost radii within discs are often empty or scarcely populated by celestial bodies, most atomic environments only slightly attenuate the intensity of open-clusters arriving from the rather broad RF range of the spectrum. It must be noted that since the absorption of open-clusters by an atom is also related to the angle under which the open-clusters has approached the atom with respect to the disc, there are always chances that an open-cluster belonging to any speed within the spectrum fails in passing a target atom. Therefore, by increasing the number of atoms on the way of RF open-clusters, chances increase that less open-clusters succeed in traversing the target environment. Attenuation of open-clusters belonging to different regions of the spectrum during their encounter with various materials is not only related to the internal dynamics of target atoms but also to target atoms' interaction with neighboring celestial entities. Unlike open-clusters forming the RF band, open-clusters forming the visible band usually end up being located on orbits already occupied by celestial bodies while traveling through most solid materials and therefore have less chances of passing through thick rows of atoms. Open-clusters forming the RF and visible bands provide examples of interaction with matter whose outcomes are mainly determined by the characteristics of target atoms. On the contrary, open-clusters forming the X-ray or gamma-ray provide examples of interaction with matter in which trajectories of open-clusters are barely affected by target atoms. In fact, these open-clusters are so fast that, once inside the galactic sphere of most target atoms, they lack sufficient time to find matching orbital trajectories corresponding to their speeds. Thus, in most of their encounters, open-clusters forming X-ray and gamma-ray force their way through galactic spheres and can mainly be stopped after collisions against massive atoms.

The arrival of open-clusters at various radii around the bulge in accordance with their traveling speeds additionally justifies the *photoelectric effect* in which electrons are ejected from a negatively charged material while being bombarded by open-clusters of sufficient speeds. After colliding with target electrons, the incoming open-clusters lose their original momentum and contingent on the dynamics of the collision may come to a halt, scatter, or decompose into celestial bodies. Obviously, out of these three possible outcomes of the encounter, only the scattering of open-clusters is detectable. Such scattering of open-clusters off the charged particles is known as *Compton scattering*. A scattered open-cluster shows a lower speed compared to its absolute value prior to the collision as a fraction of its momentum is transferred to the encountered charged particle during the encounter.

Unlike what is claimed by today's Physics under the topic of *pair production*, charged particles such as electrons and positrons cannot be created solely by photons. This is due to the fact that charged particles such as electrons or positrons are by far more massive than single open-clusters. The two electron and positron particles observed in pair-production effect are most likely generated after a momentous gamma open-cluster hits a target atom and a chunk of the atom is torn in the direction of impinging open-cluster. That is why a medium is always required in order to reproduce the pair-production effect.

Perturbing the gravitational equilibrium of galaxies beyond a certain threshold leads to the formation of open-clusters and their subsequent emission. The gravitational perturbations in materials might be initiated by means of flame, passing a current of electrons, or forcing the discs of galaxies to rapidly reorient using an alternating electric or magnetic field. The intensity of emitted open-clusters is directly proportional to the amount of celestial bodies brought within the gravitationally perturbed intergalactic space or released by the atoms forming the media. As gravitational perturbation grows in magnitude due to the increasing density of stray celestial bodies, open-clusters are formed more towards the interior orbits around the bulge, increasing the average speed of emitted open-clusters. In other words, at higher temperatures, emission's distribution-peak is blue-shifted.

Emission and absorption spectra of different elements in their gaseous states provide the most robust evidence on the discrete nature of discs around bulges. The empty regions within the emission spectra of gaseous media indicate galactic discs creating such patterns are formed by coplanar rings of celestial bodies separated by ring-shaped empty gaps or low-density celestial bodies. The absence of open-clusters of certain speeds within the emission spectrum of some gaseous environments is a demonstration that open-clusters are primarily formed and emitted from the orbits at which celestial bodies are sufficiently present. Only those rings containing celestial bodies may form and emit open-clusters while empty or low-density rings generate no open-clusters. The dense presence of celestial bodies at emitting orbits explains why material environments are opaque for the very same open-cluster speeds they emit during gravitational perturbations.

It is worth mentioning that the occurrence of red-shifts observed in the spectral lines of elements belonging to gaseous environments of faraway galaxies, is most likely due to the impact of the traveling open-clusters with stray celestial bodies on their way towards planet Earth that slows them down. Photons coming from farther celestial entities are more likely to undergo a higher number of collisions with stray pieces of celestial bodies and therefore generally exhibit larger red-shifts. Currently, Physics claims that such red-shifts are owing to galaxies recession from Earth which on its own is a consequent of the expanding Universe as postulated within Expansion theory.

Just like in gaseous media, the presence of narrow rings of celestial bodies within galactic discs separated by empty gaps is evidenced in crystalline or amorphous media. In practice, the presence of such narrow rings has been demonstrated by the emission of highly monochromatic laser beams generated from such solid media.

Because the emission speed of open-clusters is directly related to the dynamics of rotating discs in host galaxies, it can be affected by the application of magnetic or electric fields. Application of magnetic field modifies the directions of disc planes which may consequently lead to an increase or decrease in atomic discs' angular rotation velocities. Open-clusters emitted from such angular-velocity shifted discs exhibit a less or an extra momentum acquired according to the rotation direction of their parent discs. The splitting of spectral lines as a result of magnetic field application to an emitting environment is known as *Zeeman effect*. The increase or decrease in the angular velocity of discs probably occurs for those which in their tangential engagement with incoming celestial bodies, forming the magnetic lines, rotate towards the same or opposite directions, respectively. A stronger magnetic field may even lead to a higher number of spectral line splitting as the intertwined galactic discs are further pushed to their spatial movement limits. Likewise, application of an electric field to substances may affect the direction of disc planes and their angular rotation velocities. Electric fields applied to some atomic and molecular systems during emission, have been demonstrated to split or shift the spectral lines in a phenomenon known as *Stark effect*. Although, Zeeman and Stark effects were specifically discussed here with respect to the emission process, both concepts are equally valid when applied to the absorption process in matter under magnetic or electric field application. This is because both the emission and absorption of certain speed open-clusters by atoms are direct consequences of the occupancy of the corresponding orbital rings by celestial bodies within the rotating disc.

INTERACTION WITH MATTER

One of the oldest known interactions between light and matter is *reflection*. Reflection of open-clusters from a surface seems to be the result of their elastic repulsion by the net gravitational dynamics of the target medium. The elastic character of the gravitational encounter is evidenced by the equal values of the incidence and reflection angles and also by the fact that the momenta of the reflected particles are preserved after the encounter. In a reflection process, impinging small-mass open-clusters are unable to penetrate and disturb the gravitational balance of the target media. This explains why dense metallic or aqueous environments with strong internal gravitational connections between the galaxies forming their surfaces generally reflect a higher fraction of incident open-clusters compared to less dense media. Nevertheless, based on the dynamics of every encounter, there are always chances that a fraction of open-clusters dissipates into the target media.

The intricate gravitational dynamics of each medium renders its interaction with incoming open-clusters and other celestial objects unique. These unique signatures arising from distinctive gravitational interactions with incoming celestial objects are utilized to perform accurate material characterization studies such as X-ray photoelectron, Auger electron, or Infrared spectroscopies.

Another fundamental interaction between light and matter is known as *refraction*. In refraction, a beam of light suddenly deviates as it enters a new transparent environment under oblique angles. The sudden change in the trajectory of an open-cluster passing through an interface of two transparent media, under a non-perpendicular angle of incidence, occurs due to the sudden variation in the gravitational pull exerted by atoms forming the interface. Density and atomic arrangements of media forming either side of the interface are the main parameters determining the new trajectory of traversing open-clusters. At the interface, traversing open-clusters located closer to the bulge of atoms face a stronger gravitational change compared to the ones farther from the bulge. That is why, for instance, refraction angles are larger for blue-color open-clusters compared to red-color ones. The higher the difference between net gravitational pulls of the sides forming the interface, the larger the refraction angle of traversing open-cluster through the interface would be. This is because pulling force exerted by one side overwhelms the pulling force from the other side. A practical observation of refraction process is realized via obtaining the emission or absorption spectra of various elements after directing their emitted or filtered light beams to a prism which basically creates an obliquely placed air-glass interface on the way of traversing beams. Such refraction leads to the formation of an element-specific spectrum which in fact is a demonstration of open-clusters separation based on their velocities. It is worth mentioning that open-clusters belonging to X-ray and gamma-ray bands of the spectrum are much less susceptible to interface effects due to their high speeds when compared to speeds governed by galactic-disc dynamics at ambient conditions.

POLARIZATION

Assumption of disc formation in open-clusters is essential to explain polarized light's interaction with matter. Disc formation in celestial objects occurs in response to their rotational dynamics which in turn is the outcome of gravitational interactions between the constituent celestial bodies.

It is the relative orientations of disc planes between an incoming open-cluster and the target atom that determines the polarity and outcome of the ensuing gravitational encounter. An incoming open-cluster whose disc plane is approximately oriented parallel to the target atom's disc plane, under a grazing incidence, is most likely reflected after the gravitational encounter. On the other hand, an incoming open-cluster whose disc plane is perpendicularly oriented with respect to the target atom's disc plane almost certainly collides against the target atom and dissolves into it. These two distinct behaviors of open-clusters in their interaction with matter lead to what are respectively known as parallel and perpendicular polarizations of light.

The absorption fraction of open-clusters with a parallel polarization increases as the incidence angle of the beam increases from grazing to right angle at which both parallel and perpendicular polarizations show similar maximum absorption fraction values. This is due to the fact that at right angle incidences, disc plane directions for all shone open-clusters will always be placed perpendicular to the surface of target medium.

It is important to note that effects such as transmission, absorption, reflection, and refraction of open-clusters are as much about the target matter properties as they are about the polarization of incoming beams. The outcome of any interaction between light and matter largely depends on target atoms' relative disc orientations and their collective gravitational dynamics. This is clearly demonstrated by Brewster's angle whose value mainly depends on the properties of the environments forming the transparent interface. Under Brewster's angle of incidence, almost all open-clusters with perpendicular polarity are transmitted through the transparent interface.

Another interesting topic in polarization concerns what is known as the *circular polarization* of light. Circularly polarized open-clusters most likely are linearly polarized open-clusters that have obtained some degree of precession. The precession of rotating disc might be clockwise or counterclockwise which explains the corresponding circular polarization states. The fact that reflection of circularly polarized open-clusters from any target material switches their precession direction, with respect to their travel directions, is an important indication that the gravitational encounter between incoming open-clusters and target atoms is an elastic one. Otherwise, for example if the incident open-clusters were making a round-trip around the first rows of target atoms, the precession directions of open-clusters would have been maintained after the gravitational encounter leading to their reflection.

A relevant discussion here is the generation of linearly polarized open-clusters from decelerating charged particles. Intense monochromatic beams of linearly-polarized open-clusters can be generated in synchrotron or linear particle accelerator facilities. These beams provide unique opportunities especially in characterization studies of low-density sample materials. An open-cluster that is peeled off a decelerating globular-cluster tends to maintain the collective momentum of its constituting celestial bodies. Thus, one of the common characteristics of open-clusters peeled off the gravitationally perturbed globular-clusters during deceleration, in an aligning magnetic field, is their polarization direction.

It is important to note that in every accelerator, the maximum achievable speed of any charged particle is restricted by its gravitational encounters with accelerator facility's stationary constituent atoms. Thus, the limits of accelerated charged particles in such facilities are similar to those of emitted open-clusters or celestial bodies forming the magnetic field. Accordingly, the highest achievable speed of X-ray open-clusters generated in such facilities is roughly twice the speed of open-clusters forming the ultraviolet band. Hypothetically speaking, if the accelerator facility were to move in the direction of charged particle's acceleration, the achievable speed of accelerated charged-particles or open-clusters peeled from them could have further increased.

What makes the topic of charged particle accelerator facilities even more interesting is that these facilities can be used to force the collision of particles to their annihilation extent. Such collisions lead to the creation of zoos of newly formed stable and unstable celestial entities including what are termed as *quark* and *Higgs boson* particles.

SUPERCLUSTER

The broadening of a beam's cross-section as it travels, better known as beam divergence, is primarily a consequence of gravitational perturbations caused by small differences between velocity vectors of nearby traveling open-clusters. The extent of the divergence depends on the magnitude of gravitational perturbation among the traveling open-clusters or in other words on the scale of differences between their velocity vectors. A beam of sunlight, which is composed of open-clusters with almost all varieties of velocities, diverges significantly as it travels through space. On the contrary, a beam of laser which is composed of open-clusters with largely identical velocity vectors, comparably, only slightly diverges as it travels. In the latter case, neighboring open-clusters with identical velocity vectors form much larger stable colonies of open-clusters within the beam known as superclusters. The size and length of superclusters within a beam increases as velocity vectors of constituent open-clusters become more identical. In today's Physics, this property is referred to as *coherence length* and is used to indicate the degree to which collimation of certain beams is maintainable.

Another parameter that influences the gravitational perturbations and therefore the structure of superclusters in a travelling beam is the density of its constituent open-clusters. For collimated beams of open-clusters with near-identical velocity vectors, a higher density of open-clusters leads to a spatially smaller size supercluster and therefore a larger beam divergence. This is the result of increased gravitational repulsions among the nearby traveling open-clusters within denser beams.

Supercluster formation is a common character of all celestial entities including celestial bodies, globular-clusters, various galaxies, and even superclusters themselves.

DIFFRACTION

Diffraction is referred to the formation of bright and dark fringes on a screen after light, passed through a narrow opening, is shone on it. During their relatively short time passing through the opening, open-clusters that come close to border atoms of opening's frame refract towards the frame in grazing angles owing to its higher gravitational pull compared to the transparent medium of the opening. The engagement of open-clusters in the refraction process is further facilitated by the divergent character of the light beam that pushes a larger number of open-clusters towards the frame. Except for the very last few columns of atoms forming the farthest edge of the frame, refraction of open-clusters towards the frame in most cases is followed by their reflection back into the opening space. The ensued gravitational interactions between the reflected and traveling open-clusters within the opening lead to significant changes in the beam's supercluster profile. Compared to supercluster formations entering the opening, the outgoing superclusters propagate under much larger divergent angles because of the additional sideward momenta introduced by the reflected open-clusters. The refraction of open-clusters by the last few columns of atoms forming its farthest edge right before leaving the opening significantly adds to the widening of outgoing beam. The resulting equilibria between individually propagating superclusters of open-clusters determine the final layout of the diffraction pattern. Besides the geometry and material of the opening, the specifics of diffraction patterns depend on density, velocity vector, and polarization of open-clusters forming the incident beam.

Young's interference experiment is known to have provided the oldest documented observation of diffraction patterns. In this setup, light passes through two closely placed slits before forming diffraction pattern on a relatively far-placed screen. Each slit widens the incoming light into a rectangle of several diverging light stripes. The superimposition of the two rectangular beams on their way to the screen leads to a significant increase in the number of visible stripes compared to when only one single slit was utilized. This happens because the open-clusters within each stripe form new supercluster arrangements after meeting similar stripes of open-clusters under grazing angles. During this process, narrower and more collimated stripes of superclusters are formed from every two superimposing superclusters of open-clusters which are similarly distributed in space. Superimposition of the two open-cluster rays in Young's interferometer results in a large contrast between the numerous bright and dark fringes shaping the diffracted pattern.

It is worth emphasizing that polarization of the shone beam is a significant factor in deciding details of the final diffraction pattern. The gravitational interplay between two side by side travelling superclusters, each containing only a single polarization of open-clusters but perpendicular to the polarization of open-clusters in the other supercluster is minimal. The polarization direction of open-clusters is also decisive in determining whether those open-clusters which are refracted by the frame atoms are mainly absorbed or reflected back into the opening. Hence, diffraction patterns obtained from superimposition of two beams with relative perpendicular polarizations would be much less pronounced.

Open-clusters arrive at the position of slits in discrete superclusters whose specifics depend on many parameters unique to the light source. The involvement of numerous parameters makes it nearly impossible to create beams with similar supercluster distribution in space. This explains why diffraction patterns are not formed when simultaneous light beams obtained from different sources are shone separately on either of the slits in Young's interferometer. In order to obtain diffraction patterns in any type of interferometer setup, an original beam of open-clusters is used to create two or more rays using slits or beam splitters. Beam splitters, inside interferometer apparatuses, slice a fraction of the light using partially transmitting mirrors before the branched beams meet under grazing incidences, as seen in Michelson-Morley's interferometer for instance.

Diffraction is a general consequence of gravitational interactions between all celestial objects and can be observed so long as particles within the colliding beams meet under grazing angles that also requires a similar spatial distribution of superclusters between the beams. Therefore, beams of all celestial objects including electrons, atoms, and neutrons are able to create diffraction patterns, the evidence of which has experimentally been demonstrated.

ABOUT THE AUTHOR

Hiwa Modarresi was born in 1979 to a Sunni Kurdish family and has spent most of his life under Shia Sharia rule. Before leaving Iran, he obtained Bachelor of Science in *Atomic and Molecular Physics* and Master of Science in *Energy Systems Engineering* from *University of Tabriz* and *Sharif University of Technology*, respectively. He received his second Master of Science in *Nanoscience* from the Netherlands' *University of Groningen* and completed his PhD studies in *Physics* at Belgium's *KU Leuven* in 2015. Since then, the author has pursued careers in the field of *Information Technology*.

CONTACT AUTHOR

Please address your remarks on questioning or improving the textual or contextual material of the book to the following e-mail:

author@measurism.com

Thank you.

www.ingramcontent.com/pod-product-compliance
Lightning Source LLC
Chambersburg PA
CBHW020600220526
45463CB00006B/2391